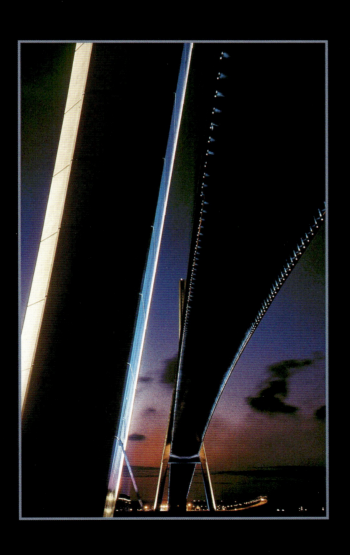

TEXT
Angìa Sassi Perino
Giorgio Faraggiana

Editor
Valeria Manferto De Fabianis

Collaborating Editor
Giada Francia

Graphic Designer
Clara Zanotti

CONTENTS

PREFACE	PAGE 10
INTRODUCTION	PAGE 12
HISTORY	PAGE 14
■ From Antiquity to the mid-18th Century	
THE IRON REVOLUTION	PAGE 24
■ From the End of the 18th to the Close of the 19th Centuries	PAGE 26
MENAI SUSPENSION BRIDGE • Great Britain	PAGE 28
PONT DE LA CAILLE • France	PAGE 30
SZÉCHENYI BRIDGE • Hungary	PAGE 32
BRITANNIA BRIDGE • Great Britain	PAGE 34
EADS BRIDGE • USA	PAGE 36
BROOKLYN BRIDGE • USA	PAGE 38
VIADUCT DE GARABIT • France	PAGE 48
TOWER BRIDGE • Great Britain	PAGE 50
FORTH RAIL BRIDGE • Great Britain	PAGE 56
PONT ALEXANDRE III • France	PAGE 62
THE ADVENT OF REINFORCED CONCRETE	PAGE 66
■ First Half of the 20th Century	PAGE 68
VICTORIA FALLS BRIDGE • Zimbabwe	PAGE 70
SALGINATOBEL BRÜCKE • Switzerland	PAGE 72
HARBOUR BRIDGE • Australia	PAGE 74
GOLDEN GATE BRIDGE • USA	PAGE 82
■ **From 1950 until Today**	PAGE 90
PUENTE MARACAIBO • Venezuela	PAGE 92
VERRAZANO NARROWS BRIDGE • USA	PAGE 96
HUMBER BRIDGE • Great Britain	PAGE 100
PUENTE LUSITANIA • Spain	PAGE 102
PUENTE LA BARQUETA • Spain	PAGE 104
PONT DE NORMANDIE • France	PAGE 108
THE BRIDGES OF THE THREE HONSHU-SHIKOKU CONNECTIONS • Japan	PAGE 116
AKASHI KAIKYO BRIDGE • Japan	PAGE 118
TATARA BRIDGE • Japan	PAGE 120
TSING MA BRIDGE • China	PAGE 122
GREAT BELT LINK • Denmark	PAGE 126
ØRESUND BRIDGE • Denmark-Sweden	PAGE 132
PUENTE LERÉZ • Spain	PAGE 136
ERASMUS BRIDGE • The Netherlands	PAGE 140
PUENTE VASCO DE GAMA • Portugal	PAGE 144
SUSPENSION BRIDGES	PAGE 148
■ Technology and aesthetics	PAGE 150
KIEL HÖRN FOOTBRIDGE • Germany	PAGE 152
SOLFERINO FOOTBRIDGE • France	PAGE 158
MILLENNIUM FOOTBRIDGE • Great Britain	PAGE 162
GATESHEAD MILLENNIUM FOOTBRIDGE • Great Britain	PAGE 166
LA MUJER FOOTBRIDGE • Argentina	PAGE 176

© 2004 White Star S.r.l.
Via C. Sassone, 22/24
13100 Vercelli, Italy
www.whitestar.it

TRANSLATION: Timothy Stroud

All rights reserved. No part of this publication may be reproduced, stored in a retrieval system or transmitted in any form or by any means, electronic, mechanical, photocopying, recording or otherwise, without written permission from the publisher.
White Star Publishers® is a registered trademark property of Edizioni White Star.

ISBN 88-544-0032-7

REPRINTS:
1 2 3 4 5 6   08 07 06 05 04
Printed in Italy
Color separation by Chiaroscuro, Turin

1 ■ PONT DE NORMANDIE - Le Havre
2-3 ■ BROOKLYN BRIDGE - New York
4-5 ■ PONT DE NORMANDIE DURING CONSTRUCTION - Le Havre
6-7 ■ TOWER BRIDGE DURING CONSTRUCTION - London
9 ■ GOLDEN GATE BRIDGE - San Francisco

11 ■ HARTMAN BRIDGE - Houston

PREFACE

When we were asked to write a book on the bridges of the world, we were rather perplexed: it was an ambitious project but for us as bridge-lovers, it was a great temptation. We considered the idea, clarifying the terms of the offer: we were not supposed to write on all the bridges in the world, but on bridges in all the world. With this clear, we were left with a fundamental problem: which bridges to choose. The most important? The best known? The most daring in engineering terms? The most beautiful? The most curious?

The selection is decidedly subjective, for which we happily accept responsibility, but it is not arbitrary: certain criteria suggested the line to take and editorial demands imposed certain limits. The work, therefore, has become an anthology of bridges. The principal criterion we adopted in this difficult selection has been that of giving the book a historical character, so we have chosen examples that demonstrate the evolutionary process of bridges during different historical epochs around the world. It should be added that the text is dedicated to bridges built during the last two centuries, with most attention paid to recent structures.

The chapter dedicated to ancient bridges provides no more than a brief description of the development of construction methods from the Roman era to the mid-18th century, but this is indispensable as an introduction to the remaining chapters.

The divisions in the text that follows were decided by the history itself of bridges. There have been particularly significant moments – in particular in the late 18th century with the institution of the École Polytechnique, the early 20th century with the invention of reinforced concrete, and the period after World War II with the application of prestressed concrete – that have opened up important new possibilities to bridge designers.

Bridges are as old as Man,

Bridges are as old as Man himself, or even older. A fallen tree-trunk over a stream, a liana between two trees on opposite sides of a river, a series of boulders that cross a stream without impeding the flow, or a rock arch eroded by the elements could all be bridges created by Nature.

Primitive man certainly used such natural crossings but went further: to join two banks, or make a ford safer, he would knock down a tree or place a flat rock on large stones set in the bed of a torrent. Such measures represented a victory of the human mind over material difficulties and, from that time, an endless struggle began. Each bridge is a response to a challenge imposed on us by Nature.

At certain times in history, the construction of a bridge has even been considered an affront to occult forces, which took their retribution by thwarting the work of man. Numerous legends have developed about how the Devil (in some cases the spirit of the river) destroyed during the night the work that had been carried out during the day, and how some bridges were completed thanks to the benevolent intervention of a god against the power of evil. In reality, the forces that oppose construction are found in the laws of physics, and can be overcome only if the laws are known. As the study of these laws gradually progressed, the daringness of the constructor increased until today, overcoming ever greater difficulties, bridges once thought impossible are constructed.

Once built, a bridge is a new presence: imposing and cumbersome, it may seem extraneous to the surrounding countryside and alter the characteristic outlines of the landscape; on the other hand, it may slip har-

14-15 ■ **PONTE DI RIALTO** - Venice
14 bottom left ■ **PONT NEUF** - Paris
14 bottom right ■ **SCULPTURAL DECORATION ON THE PONTE SANT'ANGELO** - Rome

OR PERHAPS MORE SO

moniously into the scenery and become a feature of the panorama.

Over the centuries, bridges of all types have been built. If we exclude recent cases of 'off-the-shelf' designs, devised for export, we find that rarely have architects neglected a bridge's aesthetics, but then nor have they always succeeded in making bridges attractive. At times, in an attempt to create a sense of monumentality, designers have overdone the decorations and superstructure, or have hidden the metal load-bearing structures with more 'noble' materials.

An aesthetic opinion is always subjective, but certain criteria are common. As Le Corbusier wrote, 'The false and the pompous rarely benefit art, but un-looked-for beauty can flow from the spontaneity of constructions of the engineer, which are governed by the laws of Nature and may thus achieve harmony.'

Like other infrastructures, a bridge may encourage the development of a new urban settlement or the development of zones that are peripheral to a city.

In addition to being a presence in space, bridges also represent a presence in time: History makes use of them: great events such as battles, triumphal entries, and symbolic encounters follow from their existence, but also the microhistory of the people who live near to a bridge. City bridges, in particular, are a point of attraction for the community; right from childhood, people become fond of their local bridges in the same way that people living in the countryside are fond of their mountains, plains, woods or streams.

In conclusion, bridges are singular monuments, to be used as well as admired. They are monuments built to provide a service to one and all, and, to whatever extent, they are used by one and all.

HISTORY
FROM ANTIQUITY TO THE MID-18TH CENTURY

From Antiquity to the mid-18th Century

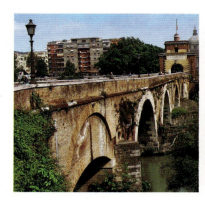 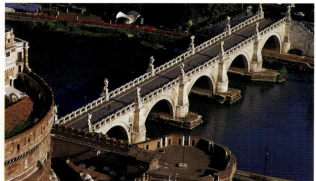

In 1570, in his treatise *Four Books of Architecture*, Andrea Palladio wrote: 'First men made wooden bridges.'...

▲ Ponte Milvio - Rome   ▲▶ Ponte Sant'Angelo - Rome

...Further on he says that only when they began to '*have regard for the immortality of their names and wealth made greater things possible did they begin to build them in stone*' (Book III). Indeed, it is personal ambition that encourages a designer to go beyond the merely functional and to produce a work of art, and it is the means made available by the client that allow him to do it. The history of bridges might begin with establishing whether slab coverings – even in their most rudimentary form – prèdated or not prehistoric wooden bridges. Of course today no trace remains of wooden bridges from the time of granite slabs, like the ancient Post Bridge in Cornwall (huge flat stones supported by piers of stacked, flat stones) or the Giang-Tung-Giao in China (with a slab that has a span of some 82 feet and a section measuring 63 by 59 inches).

A leap in quality occurred with the invention of the arch, which was known in the Middle East and Egypt in pre-Roman times, and even to the Etruscans. However, it was the Romans who took up this technique, perfected it and then applied it throughout Europe. The massive size of their empire required a secure road system that could be used in all seasons, and which therefore had to be equipped with bridges that were more solid than wooden structures. To assert their power, Rome produced monumental works not only in the *Urbs*, but also in distant provinces.

Roman bridges repeated the same characteristics: round arches, large rectangular stone blocks, great sobriety, perfection and symmetry.

Many of their bridges are still in use today thanks to many restoration operations carried out over the centuries, and, despite the occasional necessary replacement of certain elements, their original structures have not been lost and those parts that are still whole are perfectly recognizable. In Rome itself, two bridges will be discussed: the *Milvian Bridge* over the Tiber, just outside Rome, and the *Aelius Bridge*, later named *Hadrian's Bridge*, and which is today the *Ponte Sant'Angelo*, also over the Tiber but inside the city.

The MILVIAN BRIDGE was built in 109 BC and is one of the most magnificent and oldest in the entire Roman road network. It lies on the Via Flaminia at the third Roman mile outside the Flaminian Gate, now called Porta del Popolo. The monumental entrance was added during the reign of Augustus. The original length of the bridge was 165 yards and the width 25 feet. It features four slightly depressed arches, each 59 feet in span. Owing to its strategic position, the Milvian Bridge was the theater of battles and struggles: in 63 BC it was occupied by the followers of Sulla; in AD 69 it was the site of fighting between Otho and Vitellius; in 193 it saw the victory of Septimius Severus over Salvius Julianus, and in 312 the famous victory of Constantine over Maxentius; in 1849 it was where the followers of Garibaldi entered Rome; and in 1870 it saw the entry of the Italian Bersaglieri into the city.

The Aelius Bridge was the idea of the emperor, philosopher, humanist and architect Hadrian, and was completed in AD 174.

FROM ANTIQUITY

He designed it himself, and had it built to join the city to his mausoleum on the other side of the Tiber. The bridge was renovated a great deal over the centuries, both structurally and aesthetically. Of the five arches of equal span visible today, only the central three are original; the others replaced five smaller and differently designed arches that supported asymmetrical ramps. The 17th-century work done on the fortified enclosure of Castel Sant'Angelo (the fort that was built over and out of Hadrian's tomb) ended by covering much of the structure that formed the right ramp, while the new parapet designed by Bernini and the ten statues on the piers transformed the bridge into one of the most celebrated monuments of Baroque Rome.

Of the main bridges in the provinces, all of which are worthy of note, two very different ones have been chosen.

The Pont Saint-Martin over the Lys at the mouth of the Aosta valley, in Italy, was built between 70 and 40 BC. It is famous for the elegance of its structure, which rises more than 66 feet from the dry bed of the torrent. The single, very depressed arch has a span of 103 feet and a height of 37.5 feet. The ratio between the height and the span of the arch was exceptional for the period. The bridge was constructed using two different masonry techniques to confer a lovely aesthetic effect: the lower section is formed by blocks of gneiss, which are positioned without the use of binder, while the upper section is made from horizontal bands of conglomerate of flakes of gneiss and lime, and stones.

The Puente de la Alcántara over the Tagus was built in the years AD 103–104 near Alcántara, Spain, almost on the border with Portugal. The bridge, with the small temple in the square at the start of the bridge, is an example of Roman architecture at its grandest. The overall length was originally 208 yards and the width 28.25 feet. The bridge is composed of six round arches, the widest with a span of 92 feet. An honorary arch dedicated to the Roman

▲ ANJI BRIDGE - ZHAOZHOU   ▲ PONT ST. MARTIN - AOSTA   ▲ ALCÁNTARA BRIDGE - SPAIN

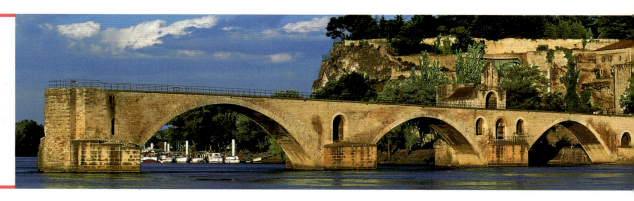

TO THE MID-18TH CENTURY

Emperor Trajan stands above the central pier. The imposing structure is made from large blocks of golden-yellow local granite with pink highlights, set together without binder. During the Reconquest of southern Spain from the Muslims (completed in 1492) by the Catholic monarchs Ferdinand and Isabella, the Alcántara Bridge suffered grave damage, but it was later restored by Emperor Charles V (1500-1558), who added the crenellation and the Hapsburg coat of arms.

During the Middle Ages few masonry bridges were built in Europe. Some were constructed in imitation of Roman bridges, and others were designed with pointed arches. In both types, it is clear that the designers acted independently of the fixed rules that had guided the Romans in their constructions; the result was that medieval bridges may have had imperfections, but they were less repetitive, more imaginative and more picturesque than their great predecessors. One of the most famous medieval bridges was the Pont D'Avignon, also known as the Pont Saint-Bénezet, built over the Rhone in Avignon between 1177 and 1185. The four arches joined to the city side of the river are all that remain of the original nineteen. Though the bridge was repaired and restored many times, floodwaters in 1680 washed the remainder of the bridge away, though the Chapelle Saint-Nicolas on the second pier remained intact and is today a wonderful example of the persistence of the ancient building systems that had been used.

Also interesting are certain ancient bridges in Asia, which demonstrate taste and high construction skills. As example is the Anji Bridge built in the 6th century AD over the Xiao river, to the southwest of Beijing. Designed by Li Chun, it has a single, very low arch built from large stone blocks without binder. The bridge was one of the large engineering works implemented in China during the Sui dynasty. The present structure has incorporated techniques such as the lightening of the gables and others that were very advanced for the era.

▲ PONT ST. BÉNÉZET - AVIGNON

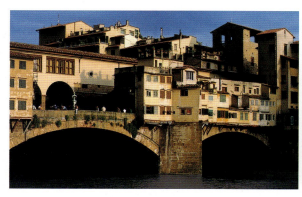

The Italian Renaissance influenced the art and style of bridge building.

▲ Ponte Vecchio - Florence ▲ Ponte di Rialto - Venice ▶ Pont Neuf - Paris

In the 15th and 16th centuries, the designers of the famous bridges in Florence, Venice and other Italian cities were inspired by the regular forms of the past, but their aspiration to be artists rather than builders led them in certain cases to be excessive in the design of the superstructures and decorations. Two examples from this period must be included.

Having survived the vicissitudes of World War II, the Ponte Vecchio over the Arno in Florence is the only ancient bridge in the city. It was built in 1345 following a design attributed to the painter Taddeo Gaddi, a pupil of Giotto. The bridge has a width of 61.3 feet and an overall length of 312 feet. Three very low arches with spans of roughly 95 feet support the superstructure within which the bridge's shops stand. This was an architectural innovation compared to other bridges of the age, in which the shops were modest wooden constructions.

Originally the superstructure was aligned with the sides of the bridge and had two filled walls with a central arch in each that allowed the public to look out over the river. The windowed backs of the shops seen today that project over the side of the bridge were built in the 16th century when ownership of the shops passed from the Commune to private individuals. Having grown fond of the picturesque appearance of 'their bridge,' the Florentines have always opposed any plans to regularize the external façade of the superstructure.

The Ponte di Rialto, Venice's famous bridge on the Grand Canal, was built between 1588 and 1591. The supporting structure is an ambitious stone arch with a span of 92 feet crowned by a magnificent superstructure. The huge construction, designed by Antonio da Ponte, has three parallel flights of steps across the bridge and two rows of shops that line the central stairway.

Fairly mediocre bridges both artistically and structurally were built in the 17th and late-18th centuries; these were for the most part banal imitations that could boast no innovations. However, one historically important bridge stands out, the Pont Neuf over the Seine in Paris. Despite its name, this is the oldest bridge in the city. It was built between 1578 and 1604 to ease communications between the Louvre and the Abbey of Saint-Germain-des-Prés. It stands beside the monument to Henri IV on the tip of the Île de la Cité and is composed of two separate structures joined by a short walk on the island. Five arches cross the left branch of the Seine and seven cross the right branch. The overall length of the bridge is 254 yards, the width is 72 feet, and the broadest span is 49 feet.

DGES

THE IRON REVOLUTION

▲ Forth Rail Bridge ▲ Eads Bridge ▲ Brooklyn Bridge

TOWARD the end of the 18th century, the French government realized that the country's engineers required a better theoretical education, with the result that the *École de Marine*, *École du Génie de Mezière* and the *École des Mines* were all set up at the *École des Ponts et Chaussées* in Paris, almost at the same time.

▲ Tower Bridge ▲ Brooklyn Bridge ▲ Pont Alexandre III

In the bridge section, a first step toward professionalizing the engineers had been taken in 1716 with the creation of the *Corps des Ponts et Chaussées*, within which was the *Bureau des Dessinateurs*. In 1747 Jean-Rodolphe Perronet replaced the *Bureau* with the *École des Ponts et Chaussées* and in 1795 the *École Polytechnique* was instituted to provide techniques for dealing with practical problems. From that time, techniques became applied sciences: the cut of the stone was given rules and procedures suggested by descriptive geometry, and treatises ushered in scientific 'stereotomy,' which was defined as 'the art of making use of the weight of stone against itself and of using the stone's weight to support itself.' The construction of the vault had passed from being a practical problem to a scientific question. The Age of Reason, transformed 'the art of fabricating' into the 'science of building.' The planning and building techniques related to bridges underwent a leap forward in quality: having developed a 'manner' founded on theory and tested in numerous applications, Perronet inaugurated a new era in the construction of stone bridges. 'Perronet bridges,' represented the best type of stone bridge, later replaced by the masonry bridge. While France was setting up its prestigious technical and scientific schools, in England engineers got their training by working in workshops and building sites. They called in mathematicians and physicians to help on tangible problems, but if the answer of the scientists did not convince them, they turned to building models and prototypes. The Industrial Revolution led to the use of ferrous materials and extended their use to civil and industrial construction. Between 1776 and 1779, the first cast-iron bridge was built over the River Severn at Coalbrookdale, and, in 1796, a bridge was erected over the River Wear between Sunderland and Monkwearmouth. Although these first attempts made use of the new material without exploiting its specific characteristics, they were at any rate a starting point and they stimulated the study of the new bridge type. The 19th century, in particular the second half with the development of the railroads, was an exciting period for bridges. The best of the art and science of construction was concentrated on the railroad bridge in the 1800s: it became a testing ground for structural theory and was the most effective form of application.

From the End of the 18th to the Close of the 19th Centuries

28-29 ■ The Menai Bridge between Wales and the Isle of Anglesey crosses the Menai Strait.

29 top right ■ Workers replacing parts in the chain in 1939, an operation that has needed to be carried out several times.

29 bottom ■ Shown here is the stone structure built to stabilize the towers, which are subjected to horizontal thrust by the transit of vehicles.

Menai Suspension Bridge

Menai Strait, between Wales and Anglesey Island (Great Britain)

In ancient times the Menai Strait had always acted as a natural defense against the ambitions of conquerors, but more recently it became an impediment to communications. In 1801, when Ireland was united politically with England, the connection between Wales and the island of Anglesey, which for centuries had been by ferry, no longer seemed adequate. The Strait lay in the path of the postal route between London, the port of Holyhead and Dublin. The debate on finding an alternative to the ferries, which had been going on for 20 years, flared up once more but without achieving anything. The political situation at the time of the American Revolution and Napoleonic wars required heavy economic commitments and did not permit costly projects to be undertaken.

In 1818 the Menai Strait question once more came to the forefront, and it was decided that a bridge would be built that would be at 100 feet above the high tide level to allow sailing vessels to pass. The engineer Thomas Telford's design for a suspension bridge met this requirement, and his plan was immediately approved. The first stone of the structure was laid in August 1819. At that time, the suspension bridge was a state-of-the-art solution that had been adopted in only a few instances, each of which had represented an occasion to experiment with new systems.

Thomas Telford (1757-1834) was one of those British engineers who called themselves 'practical men,' but they did not disdain making use of the consultancy of theoretical engineers and mathematicians. In 1820, when already known and admired as a builder of large public works, Telford was appointed the first president of the

Location	Designed by	Length – Widest Span	Type	Construction
Wales (Great Britain)	Thomas Telford	1,709 ft / 579 ft	Chain Suspension Bridge	1819-1826

Institution of Civil Engineers. With the construction of the Menai Bridge, Telford showed that he was able to reconcile technical problems with aesthetic and budgetary requirements.

The bridge has a suspended span of 579 feet between two stone sections; the one on the Anglesey end of the bridge is formed by four arches and the one on the Welsh end by three arches. The inner ends of the stone sections are where the two stone towers stand that support the four suspension chains. Each chain – composed of four flat sections connected and hinged together – was made from puddled iron, a material that became reliable only after 1850. With the assistance of Samuel Brown, Telford obviated any possibility of failure by carrying out the necessary load tests to ensure the mechanical characteristics of the iron were adequate; the results were excellent and the chains, with a little restoration and occasional replacement of links, lasted until 1940 when they were replaced by identical steel versions. The deck, originally made of wood but replaced with iron in 1893, was connected to the chains by 444 ties. The task of positioning the chains was adventurous and spectacular. For each of the four chains, a section weighing over 20 tons was floated on a raft beneath the central span, and another section positioned on either side and anchored in both banks. The side sections were then made to pass over the top of their corresponding tower and allowed to dangle over the water. The final operation was to match the sections up and join their ends. This job required 150 men, who had only hand-operated winches to help them. Accompanied and encouraged by a band of pipers and drummers, the navvies worked before an enthralled crowd, which applauded enthusiastically at the end of the job. The bridge was completed in 1826.

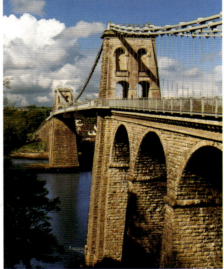

Pont De La Caille

over the Les Usses torrent, Allonzier-la Caille and Cruselles, Haute-Savoie (France)

Until 1856 Savoy was one of the Sardinian States. The region was connected to Switzerland by the *Route royale de Chambéry à Geneve par Annecy* which crossed Les Usses torrent at the bottom of the deep Gorge de la Caille (Quail Gorge) where there was a Roman bridge. Using the technique that had become established in the 1830s, it was possible to build the suspension bridge that is still in use. It crosses the torrent in a single span of 629 feet at a height of 482 feet. It was called *Pont Charles-Albert* in honor of the reigning prince, Charles-Albert of Savoy, king of Sardinia. The bridge is known today as the *Pont de la Caille* or *Pont d'Annecy*.

The bridge was inaugurated on October 7, 1839 in the presence of the king. At the time the bridge had one of the largest spans in existence, and it aroused admiration for its technical qualities, aesthetics and its dramatic natural setting.

The support piers for the chains are built in Neo-Gothic style: they are a pair of cylindrical masonry columns 57 feet high, joined by a wall through which an arch passes. The suspension cables pass through a large aperture in the columns.

The persons responsible for the design and construction have not been definitively

Location	Designed by	Length – Widest Span	Type	Construction
Haute-Savoie (France)	F.P. Lehaître	761 ft / 630 ft	Suspension bridge	1839

30-31 and 31 bottom ■ Pont de la Caille crosses the gorge created by the Usses torrent near Annecy. The cable anchors have devices that allowed the cables to be tightened.

identified as the archival sources do not agree. The initiative to build the bridge was certainly taken by the Société Bonnardet et Blanc, which had the concession, but in some documents the name of F. P. Lehaître appears as the only designer as well as director of the works, and in others the same responsibilities are attributed to the engineers Ch. Berthier, E. Bertin and the *Inspecteur des Ponts et Chaussées*, Monsieur E. Belin. Since 1928 the old suspension bridge has been relieved of motor traffic. In that year, a short distance away, a reinforced-concrete arch bridge was opened, the Pont Caquot. It has a span of 451 feet and height of 87 feet.

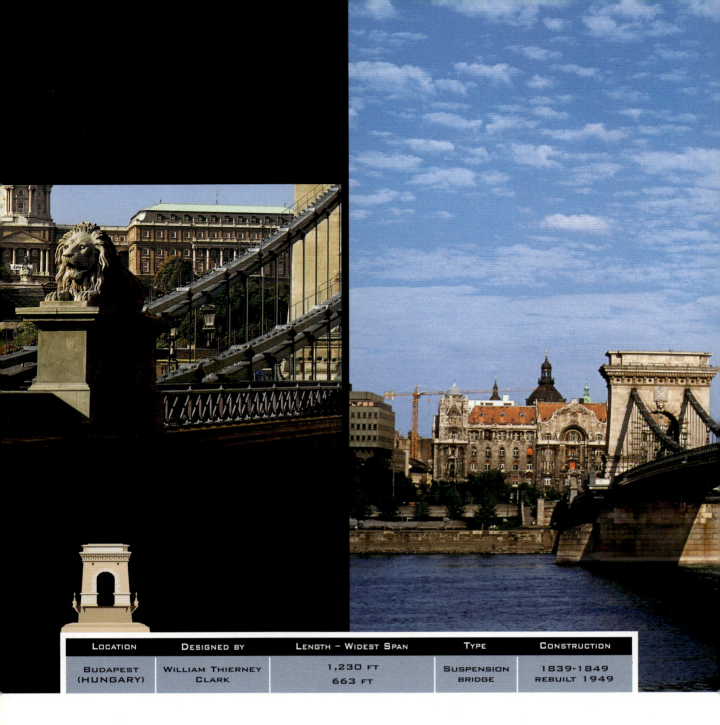

Location	Designed by	Length – Widest Span	Type	Construction
Budapest (Hungary)	William Thierney Clark	1,230 ft / 663 ft	Suspension Bridge	1839-1849 rebuilt 1949

Széchenyi Bridge

Danube river, Budapest (Hungary)

The Széchenyi Bridge, also known as the Chain Bridge, is strongly tied to the history of Hungary and Budapest in particular. Construction lasted ten years, from 1839 to 1849, having been begun during the era in which the country was governed by moderate reformists, and continued during the period of the Hungarian uprising that culminated in 1849.

The historical significance and symbolic value that the Chain Bridge has for Budapest are such that the bridge, which was destroyed during World War II, was rebuilt as a faithful reproduction of the original.

The Széchenyi Bridge is named after the statesman who championed its construction, but outside the country it is known simply as the Chain Bridge for a very rare, though not unique, feature: the deck is not suspended from cables but from chains.

The designer was the renowned English engineer William Thierney Clark (Bristol 1733-London 1852), who was an expert bridge designer and constructor.

The suspension bridge has three spans and is 1230 feet long overall; at 663 feet the central span was the world's longest at the time, and it was only with the Brooklyn Bridge in 1883 that suspension bridges doubled the previous spans.

The chains – two on each side, pass through apertures just below the cornice of the towers. The distance between them is 26 feet,

32 ■ The ends of the bridge are guarded by four majestic stone statues of tongueless lions.

32-33 ■ The Chain Bridge was the first permanent bridge over the lower stretch of the Danube, and was built to join Buda and Pest.

33 bottom ■ The pedestrian and cycle lanes are on either side of the vehicle section, with balconies at the base of the arches.

which limits the width of the bridge. The chains are anchored at either end of the bridge in blocks of stone that resemble the piers.

In 1914 several structural elements needed replacing because of wear over sixty years of use, and modifications were required to adapt to the new traffic needs. Later, the military actions in Budapest during World War II resulted in the destruction of all its road and railway bridges, including the Chain Bridge.

Responsible for the unity of the two cities Buda and Pest, and part of the urban landscape for almost a century, the bridge had to be rebuilt. So, despite the postwar difficulties, reconstruction began immediately and in 1949 the new but faithful reproduction of the previous bridge was opened.

The Chain Bridge reflects the eclectic – or rather the historicist – canons of the era in which it was designed: the ashlar-work, the cornice and the proportions of the open arches in the piers were inspired by Italian Renaissance architecture.

Britannia Bridge

Menai Strait, between Wales and Anglesey Island
(Great Britain)

34 ■ The original bridge, with large lions at the entrances, seen left in a painting in the National Railway Museum, had a twin-tunnel truss for trains.

35 top ■ In 1849 readers of the *Illustrated London News* could follow the construction of Britannia Bridge through the drawings it printed. The enormous iron trusses were built on the river bank, then transported by flatboat and raised to their final positions, with the ends resting on the two towers.

35 bottom ■ After the fire in 1970, the bridge underwent renovation of the springers: the old caisson girders with an arched supporting structure were replaced and two levels were built on an arch truss.

The development of the railroads in Britain was exceptionally rapid and by the 1840s required the construction of adequate bridges. One of these was the Britannia Bridge over the Menai Strait, about a mile away from the suspension bridge built by Thomas Telford in 1826.

The Britannia Bridge was named after the tiny islet on which the central tower stood, and it can be considered the precursor of the light metal structure.

The designer, Robert Stephenson (the son of the famous inventor of the locomotive), worked with the metal engineer William Fairbairn, a well-known shipbuilder.

In order not to halt the shipping in the strait, the possibility of an arch bridge was ruled out as the support cooms would have blocked the passage of the ships during construction period. Today the bridge is very different from the original one.

In 1970 a fire broke out, caused by a group of irresponsible people who lit some sheets of paper so as to see better in the darkness of the tunnel.

The bridge was damaged throughout, and new support archways had to be built to replace the original girders. At the same time, a road deck was constructed above the rail-

Location	Designed by	Length – Widest Span	Type	Construction
Wales (Great Britain)	Robert Stephenson and William Fairbarn	1,512 ft / 479 ft	Metal bridge with tubular trusses	1846-1850

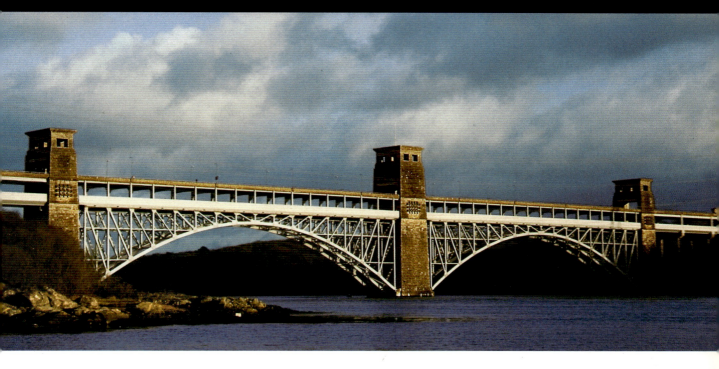

road level, which today has just one set of tracks.

What remains of Stephenson and Fairbairn's original structure is a continuous truss of four bays formed by two tubular girders side by side, so large (30 feet high by 16 wide) that a train can pass through each.

The five piers are made from stone quarried in nearby Penmon, and the three middle ones are topped by towers 220 feet high. A gigantic statue of a lion on a stone base, sculpted by John Thomas, adorns each of the two entrances to the bridge.

The towers are merely there for aesthetic reasons, but at the start they were designed to support the chains that should have contributed to support the tubular truss (in which holes were prepared *ad hoc* to take the chains).

However, during construction, Fairbanks realized that the truss could take the loads without the use of supplementary chains. He had intuitively understood the behavior of the continuous structure, which was only recognized and demonstrated theoretically at a later date. Confident of his hunch, he prepared some tests and, encouraged by the positive results, decided to leave the chains out.

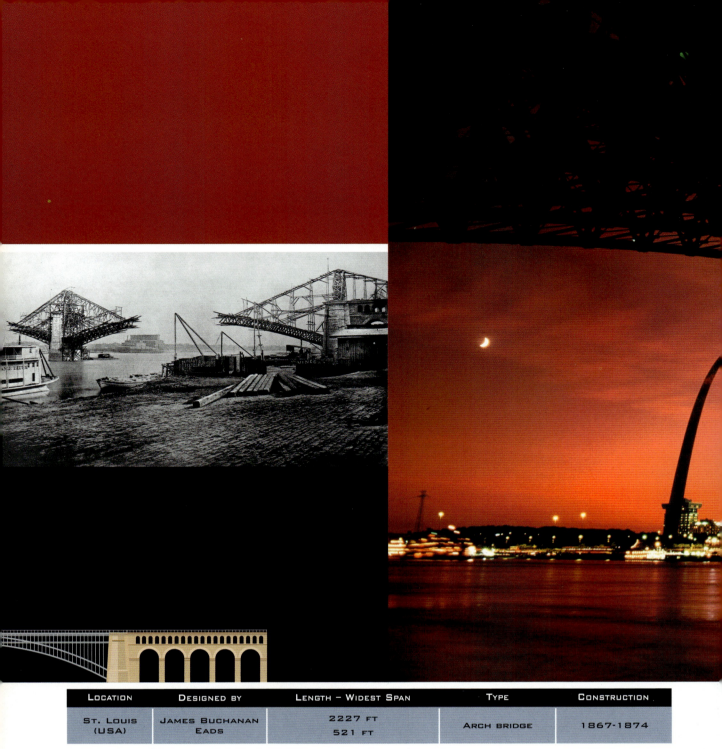

Location	Designed by	Length – Widest Span	Type	Construction
St. Louis (USA)	James Buchanan Eads	2227 ft / 521 ft	Arch bridge	1867–1874

Eads Bridge

Mississippi River, St. Louis, Missouri (USA)

In the United States the railroads had covered most of the country by the middle of the 19th century. They greatly stimulated trade, particularly in the largest cities. However, St. Louis did not draw the same benefits provided by the railroads as did Chicago until a connection between Missouri and Illinois was created. This required the construction of a bridge over the Mississippi. In 1867, two years after the end of the Civil War, construction of the bridge was begun by the Keystone Bridge Company. The designer and director of the works was James Buchanan Eads (1820–87), a self-taught American engineer known for having invented, as a young man, a type of flatboat for recovering sunken steamboats in the Mississippi, and in 1861, in response to an order from the Federal government, building eight armored steamboats in just 100 days. In constructing the bridge in St. Louis, Eads tried out new construction methods that he soon patented. The load-bearing structure of the Eads Bridge is made of steel. It has three latticework arches with tubular elements; the spans of the bays were exceptional for the era, with the central one measuring almost 525 feet. The arches support a reticular latticework tunnel through which the railroad runs, and a paved road deck above. On either side of the river, the bridge continues beyond the banks on stone structures, each supported by four small arches. In order not to interrupt the river traffic during construction of the bridge, Eads dreamed up the system, hitherto

36 ■ In 1871 the arched spans were built using a new cantilever system that allowed river traffic to pass unhindered.

36-37 ■ The high, slender Gateway Arch, the new symbol of St. Louis, is framed beneath one of the arches of the historic Eads Bridge.

37 bottom ■ Traditional, multistory riverboats pass under the arches of the bridge.

unknown, of cantilevering. When the arches were finally closed, the measurement of the bars was inexact. Eads overcame the difficulty by cutting the tubes and inserting a threaded connector so as to be able to regulate the length during assembly; this expedient also provided the possibility of dismantling and replacing pieces. Eads realized this and transformed what had been a dodge into an innovation he could patent. He used compressed-air caissons to sink the foundations of the piers, but at that time the effects on men of rapid decompression (the 'bends') were unknown. For this reason, he used the new method of underwater excavation without taking the slightest precaution and with a great deal of cynicism. During construction of the Eads Bridge, twelve men died.

The bridge was tested by subjecting it to a weight of 550 tons in movement, i.e., a weight equal to that of 14 locomotives Subsequently, a wind-resistance trial occurred naturally during the great tornado of 1896. In 1991, after more than a century of use, the bridge was closed. In 1993 it was reopened but only for light railroad transit. Today, after more than 12 years of abandonment, the Eads Bridge has been reopened to traffic. On the south side of the bridge, which gives an excellent view of the Mississippi and the city of St. Louis, there is a cyclist and pedestrian lane, which has been greatly appreciated since the restoration of the bridge.

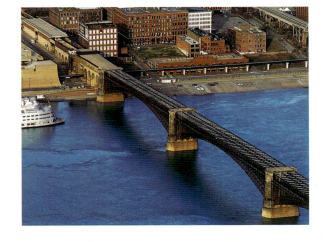

Brooklyn Bridge

East River, New York (USA)

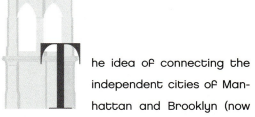

The idea of connecting the independent cities of Manhattan and Brooklyn (now boroughs of New York City) had been discussed since 1806. Studies were carried out on the feasibility of the project, which included a proposal for a tunnel as, at that time, a tunnel appeared to be less difficult to construct than a surface crossing. After more than 60 years of sometimes acrimonious debate, events moved forward. In 1869 the design that John Augustus Roebling had presented to the New York Bridge Company on September 1, 1867 was finally approved. The introduction to the *Plan and Details of Anchorage, Approach Towers, and Steel Cables* says: *'The contemplated work, when constructed in accordance with my design, will not only be the greatest bridge in existence, but it will be the great engineering work of the Continent and of the age. Its most conspicuous feature – the great towers – will serve as landmarks to the adjoining cities, and they will be entitled to be ranked as national monuments. As a great work of art, and a successful specimen of advanced bridge engineering, the structure will forever testify to the energy, enterprise, and wealth of that community which shall secure its erection.'* With these words the author declared immodestly the ambitiousness of his project and the certainty that it would be realized. John Augustus Roebling

38 ■ Brooklyn Bridge crosses the East River southwest from Manhattan to Brooklyn. The aerial view shows the vehicle lanes, partially hidden by the transversal trusses of the covering.

39 ■ The massive stone portal and Gothic arch is one of the best known and most representative images of the bridge. The diagonal stay cables were added to the vertical suspension cables to give the platform greater stability.

40 top ■ The base of the piers has to be very solid and stable: this drawing from 1876 shows a section with the caisson used for the excavation work.

40 bottom ■ Men inside the caisson worked in a pressurized environment and the material excavated was raised up the central column.

40-41 and 41 bottom ■ A foggy Manhattan provides the backdrop to two workers in an acrobatic and dangerous position. It is 1875. The two towers had been completed and the four suspension cables tensed. Construction then began on the lower section of the deck, from both sides, and was completed in 1883.

(1806–69) had acquired a solid theoretical education in Germany where, in 1826, he had graduated in Civil Engineering at the Royal Polytechnic Institute in Berlin. In the United States, where he had arrived as an immigrant in 1831, he acquired professional experience at high level: he built important works such as the Allegheny Aqueduct over the Allegheny River, the Monongahela Bridge in Pittsburgh, the Delaware Aqueduct Bridge – still in existence – and the 352-yard Ohio Suspension Bridge in Cincinnati. At the end of the 1860s, the city of New York was experiencing great expansion; in the previous decade the population had risen from 266,000 to 396,000 inhabitants with a percentage increase greater than that of any other American city. Business with the city of Brooklyn had also grown enormously and the construction of a bridge had become an increasingly urgent need. The City Council,

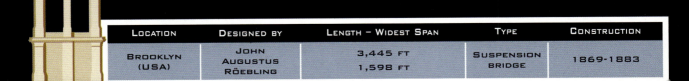

Location	Designed by	Length – Widest Span	Type	Construction
Brooklyn (USA)	John Augustus Röebling	3,445 ft 1,598 ft	Suspension bridge	1869-1883

42-43 In addition to the Brooklyn Bridge, the Manhattan, Williamsburg, and Queensboro bridges, and two tunnels connect Manhattan to Brooklyn and Queens.

43 bottom The cycle lane is in the central space between two suspension cables on the upper level.

which in 1857 had allocated funds for that very project, approved the costs in 1866 and chose Roebling's innovative design. In his calculations for the structure Roebling contemplated the use of steel, a material that was still little used, as it had a resistance that was double that of common iron. Even the building equipment was completely innovative: for the first time compressed-air caissons were used during excavation of the foundations beneath the water level. Unfortunately, construction of the bridge was hindered by many unpleasant episodes. First was the death of Roebling himself, following an accident. Before the start of the project, Roebling had had a foot crushed on a ferry during an inspection of the pier positions. Tetanus set in, and he died just a few days later, on July 20, 1869. Responsibility for direction of the project passed to his son Washington, who had acquired the necessary experience working alongside his father during construction of the Ohio Suspension Bridge in Cincinnati. In order to follow the excavation of the foundations beneath the water level in person, in 1872 Washington Roebling entered the compressed-air caisson and was affected by decompression syndrome ('the bends').

44-45 ■ Every day some 140.000 vehicles cross the bridge, a number that gives a very different impression to this unusual sight.

44 bottom ■ The sun creates contrasting effects between the cables and the double portal in shadow in the background.

46-47 ■ The Statue of Liberty can be seen between the vertical suspension cables and the oblique stays.

At that time the cause of this condition was little understood: many workers remained paralyzed as a result of working in compressed-air conditions, others lost their lives. On this occasion, Washington Roebling, who became paralyzed, suspected the cause and decided not to continue on the excavation, which however had reached a sufficient depth. He was obliged to continue direction of the project from a window in his house. Construction of the bridge took 14 years, during which time many other accidents occurred with subsequent loss of life. On May 23, 1883, the Brooklyn Bridge was finally inaugurated: at 1,600 feet long, the main span was the largest ever built, a record it retained until 1903. The appearance of the Brooklyn Bridge is known the world over: the deck is a reticular metal structure over 1,100 yards long and hangs from four cables anchored at the ends and supported by two granite Neo-Gothic towers. These stand 1,600 feet apart and rise 325 feet above the water. The bridge's two parallel roads, each of three lanes, pass through pointed arches in the towers. Today the bridge provides passage across the East River for road traffic, light railroad traffic and pedestrians.

48 ■ The light and rigid structure of the Garabit viaduct offers no resistance to the wind and could be built in a very short period.

48-49 ■ The central arch becomes slimmer toward the ends until it joins two very wide hinges that provide lateral stability.

49 bottom ■ Alexandre-Gustave Eiffel, primarily a bridge builder, subsequently became famous for the Eiffel Tower and work on the Statue of Liberty.

Viaduct de Garabit

Truyère River, near Saint-Flour, Cantal (France)

In 1878 the path of a new railroad was approved between Béziers and Clermont-Ferrand. Two possibilities were considered, and the one that seemed generally more practical was chosen. However, it required overcoming one very large technical obstacle: the crossing of a deep gorge in the Truyère valley. Though problematic, this route was selected because a previous construction (an arch Alexandre-Gustave Eiffel had built over the Douro river at Porto, in Portugal) suggested that however ambitious the project was, it was not impossible. The administration believed, though, that only Eiffel was capable of such a challenge and it commissioned him with the task directly, thus violating the rules of the public tender. The name of Gustave Eiffel (Dijon 1832–Paris 1923) was not only known for the bridge over the Douro, but also for when, ten years before, he had published an important study on materials used in metal constructions and was the owner of the company *Société de Constructions de Levellois-Perret*. The company (which was later to construct the famous Eiffel Tower in Paris) had already worked on many important engineering works, including the metal viaducts over the Sioul and Neuvial rivers.

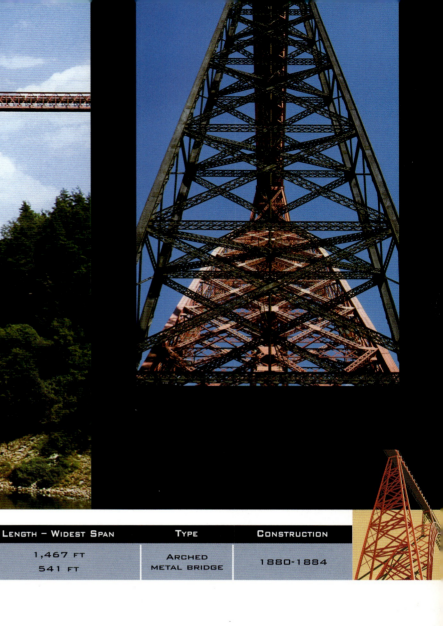

Location	Designed by	Length – Widest Span	Type	Construction
Saint-Flour, Cantal (France)	Alexandre-Gustave Eiffel	1,467 ft / 541 ft	Arched metal bridge	1880-1884

The Garabit Viaduct is a metal reticular structure that crosses the valley at a height of more than 400 feet in a huge scythe-shaped arc (541 feet span and 171 feet height). The deck is 1,467 feet long and is formed by independent trusses that can withstand the deformations of the structure caused by the passage of a train. The viaduct stands on piers of different height, in the form of truncated pyramids made from a wrought-iron latticework.

The tallest piers (394 feet) reach the shoulders of the arch. The central truss is supported by the arch; it rests directly on it at the center-point and on two piers on either side. The paths of the stresses are clearly visible in this apparently light and elegant bridge. Its innovative design in comparison to the traditional masonry or wooden structures made it one of the most famous engineering works of the second half of the 19th century.

TOWER BRIDGE

RIVER THAMES, LONDON (GREAT BRITAIN)

The need to build a new bridge in the East End of London was recognized from the middle of the 1860s. A decade later the question had become urgent because the strong increase in traffic over London Bridge had become excessive (in 1882 the bridge carried a daily average of 22,242 vehicles and 110,525 pedestrians). However, the idea of a crossing downstream was negated by the port activity because a bridge near the port would have blocked the passage of the ships.

The problem was finally resolved by Sir Horace Jones when he presented an architectural project for a bascule bridge in which the road level, divided in two sections transversally, could be raised by rotating each section around two pivots inside the bridge so as to allow ships taller than the clearance to pass through. The design was approved and brought to fruition with the calculations of Sir John Wolfe-Barry.

The famous Tower Bridge comprises three spans divided by two large towers 207 feet high. The central span can be raised hydraulically, whereas the two lateral spans are suspended from inverted reticular arches. At either end of the bridge there are smaller towers, 143 feet tall, to which the suspension trusses are connected. At a height of 142 feet,

50 ■ The bridge looks like a medieval castle but hides its iron framework beneath stone cladding. The towers contain the mechanism that opens the bridge.

51 ■ Tower Bridge is the first bridge over the Thames as one arrives from the sea. It was designed to open to allow ships to pass through to reach the port area. The Tower of London can be seen at the top on the left bank.

the larger central towers are joined by two reticular structures that have the function of absorbing the strain imposed by the suspension bars and providing a pedestrian walkway. The towers have metal frames lined with Cornish granite and Portland stone.

The mechanism that raises and lowers the two bascules is housed in the two larger towers. The lifting system was innovative at the time and is formed by a hydraulic mechanism and an enormous toothed wheel with a pinion gear.

The death of Horace Jones in 1887 left responsibility for the works totally in the hands of Wolfe-Barry, who was free to make changes to the design, including those of an aesthetic nature. At the end of the works Londoners realized that the bridge was no longer the one designed by Jones. With its Victorian towers, it was far from the Neo-Gothic style architecture seen in the original project.

52-53 ■ A worker oils the motor that keeps the hydraulic accumulators under pressure, which enable the bridge to rise.

52 bottom ■ The first ideas for the bridge, designed by Sir Horace Jones, were for an arch in the central span.

53 top ■ Construction of the bridge did not impede river traffic that had to reach the port.

53 bottom ■ The opening ceremony took place on June 30, 1884 with the participation of the royal family.

Location	Designed by	Length – Widest Span	Type	Construction
London (Great Britain)	Sir Horace Jones and Sir John Wolfe-Barry	940 ft 269 ft	Bascule bridge	1886-1894

Tower Bridge

54-55 ■ The truss and parapet decorations feature the strongly contrasting colors of red, white and different shades of blue.

54 bottom ■ The upper walkways allow pedestrians to cross even when the bridge is open. They also have a static function as tie-rods.

There were protests from the public and the press did not restrict its criticism.

But with its impeccable operation, the bridge defended itself. The two bascules, each weighing 1,000 tons, opened and closed in less than a minute and a half. They functioned regularly and continued to do so uninterruptedly for many decades. The results rehabilitated Sir John Wolfe-Barry and his reputation was saved.

Tower Bridge soon became – and still is – a symbol of the City of London as well as being a tourist attraction. Each day 150,000 vehicles cross the bridge, and the river traffic, though reduced, requires the bridge to be opened about 500 times a year.

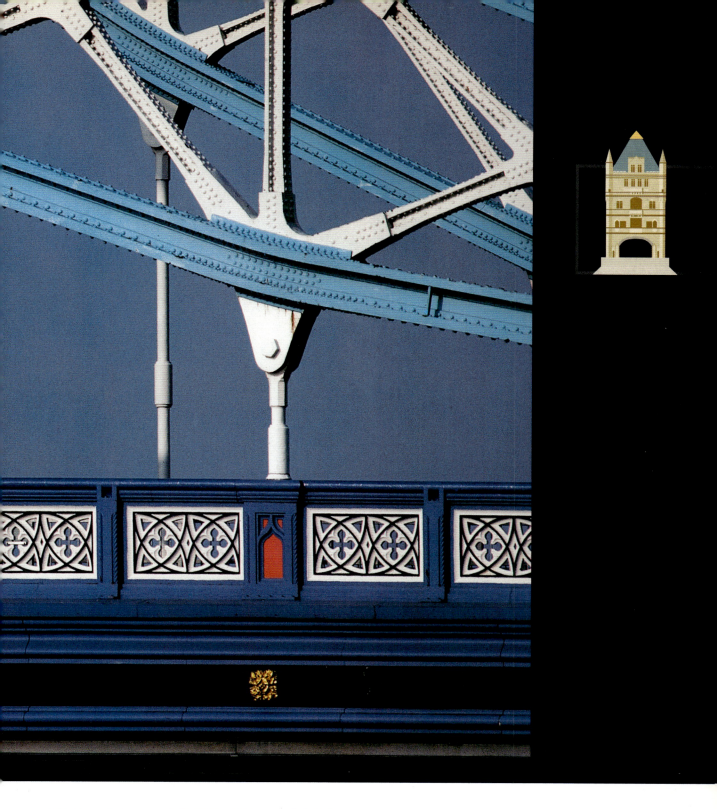

TOWER BRIDGE

55 bottom left ■ The external towers are lower than the central ones and hold up the ends of the upturned arches that support the spans.

55 bottom right ■ The two moving sections of the bridge have heavy counterweights inside the towers.

Forth Rail Bridge

Firth of Forth, near Edinburgh (Great Britain)

The Forth Rail Bridge was designed and built by John Fowler (1817–98) and Benjamin Baker (1840–1907), two engineers of fame who were known for having previously directed important projects for the British government. Because the bridge was built shortly after the collapse of the bridge over the nearby River Tay, it was natural that the two engineers should be particularly careful in their design and calculations. They chose an isostatic structure (i.e., one indifferent to yield and temperature variations and that could withstand construction errors); they engineered the construction to resist the worst possible conditions and included extreme safety coefficients. The result was a heavy and costly structure but it was – and still is – admired for its imposing presence and solidity.

The structure is a gerber type (cantilever), named after Enrico Gerber who tested this type of truss on several large bridges in Germany. The Forth Bridge has three gigantic trusses 361 feet in height that rest on square granite bases. Each truss supports two symmetrical brackets 681 feet long. At the ends of the brackets are the central trusses, each 342 feet long, making each span 1,704 feet in length. The bridge extends over both sides of the firth, with a continuous truss that rests on granite pylons.

Overall the metal surfaces of this gigantic bridge are equal to 180,000 square feet. The compressed lower bars are tubes with a diameter of 11.75 feet. The span of the bays was

56 ■ The Forth Bridge is one of the most solid structures ever made. It has exceptionally wide bases to prevent the wind from blowing the structure over sideways, as happened to the nearby Tay Bridge.

57 ■ The road suspension bridge was built next to the old railway bridge in 1964, slightly upstream in the Forth estuary.

a record for that type of structure and was only exceeded in 1917, when the slightly larger bridge over the St. Lawrence River was built in Quebec.

The gerber truss was effectively demonstrated when Kaichi Watanabe, Baker's assistant, used his own weight to correspond to the load that the central bay placed on one end of the two brackets. The two men seated with their arms out (representing the tie bars) supported two rigid elements (struts) that corresponded to the elements compressed by the brackets. Each of the weights suspended at the sides (equal to half of the central weight) represented the action of the adjacent truss that weighed on the other end of each bracket.

The construction of the Forth Rail Bridge cost the lives of many workers, something that occurred too often but which in that era was accepted as a natural consequence of bridge building. The bridge was inaugurated on March 4, 1890.

Location	Designed by	Length – Widest Span	Type	Construction
Edinburgh (Great Britain)	John Fowler and Benjamin Baker	8,202 ft / 1,709 ft	Gerber metal truss	1883-1890

58 top ■ Construction of the structures progressed simultaneously and symmetrically from the three bases until they met in the middle.

58 bottom and 59 center and bottom ■ The 3 drawings, copied directly from the bridge by Baker's assistants in 1890, show the structure of the construction.

58-59 ■ Painting and maintenance are performed continually and require personnel that do not suffer from vertigo.

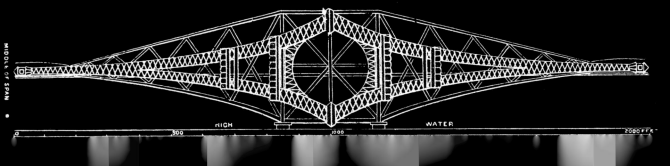

FORTH RAIL BRIDGE

60-61 top ■ The structure is a complex network of struts tensed and compressed by the bridge's enormous dimensions. The weight and volume of the train are negligible in comparison to that of the bridge.

60-61 bottom ■ Sailboats seem out of scale when seen against the bridge. During sunset, the reddish color of the spans seems to seep into the sky.

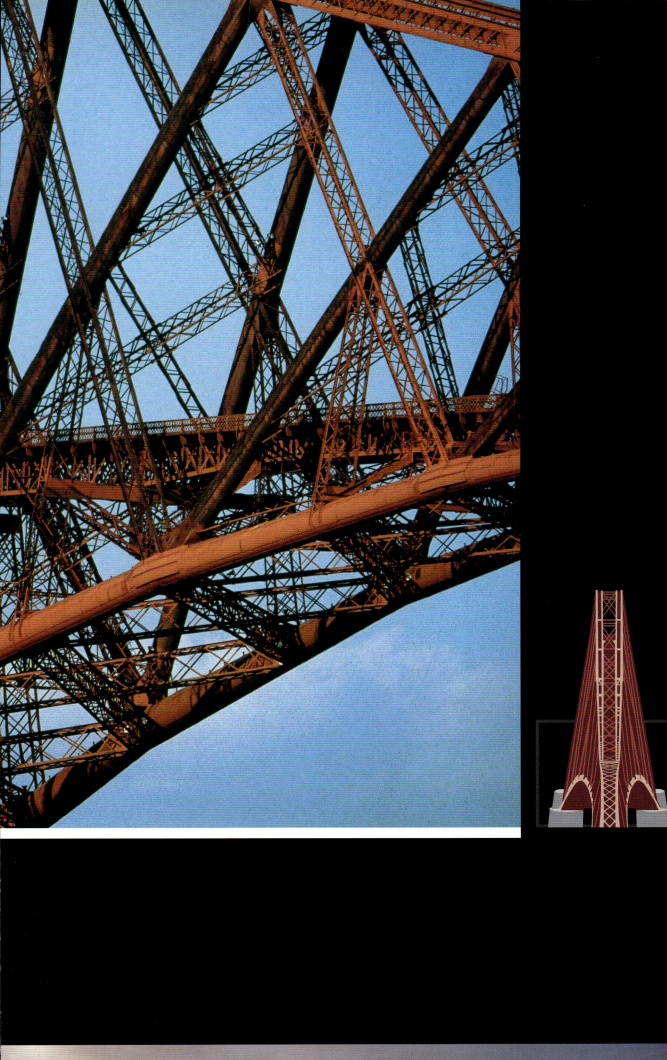

Pont Alexandre III

River Seine, Paris (France)

The Alexandre III Bridge was built as part of the 1900 Exposition Universelle to connect the pavilions built on the left bank to the palaces dedicated to the Beaux Arts on the other. It was decided that the bridge also had to be decorative and imposing, but at the same time that it should not spoil the view toward the set of monuments known as Les Invalides; another problem was that river traffic would be impeded if a bridge pier was positioned there (just downstream of a bend in the river and immediately upstream of the Pont des Invalides). Therefore it was necessary to build a single-arch bridge. The engineers Jean Résal and Amédée Alby, with the assistance of the architects Bernard Cassiet and Gaston Cousin, presented a design that reconciled all the various requisites. On October 7, 1896, during a

62-63 ■ A taste for abundant decoration distinguishes many bridges built in the late 19th and early 20th centuries, but on the Alexandre III Bridge it exceeds every expectation.

ceremony dedicated to the recent friendship pact signed between France and Russia, Tsar Nicholas I placed the first stone of the bridge, which was named after the recently deceased Tsar Alexander III.

Construction of the bridge began in early 1897 and was completed in time to be opened before the inauguration of the World Fair.

The bridge crosses the Seine with a low single-arch span of 351 feet. The structure is formed by fifteen parallel three-hinge arches built from forged steel quoins that were assembled on site. The deck is made from laminated steel and is 131 feet wide.

What is most striking about the bridge, other than its exceptional width, is the abundance of decoration that took a trend of the time to the extreme. Even at the design stage, the amount of decoration was evident, as there are two large masonry pylons built at either end as bases for massive statues. The gables of the bridge too are heavily adorned with statues and embellishments, and the extraordinary richness of the superstructure was the work of many artists. The two bronze statues of *Nymphs of the Seine* and *Nymphs of the Neva* (on the keystones on either side) are by Georges Récipon, and the allegorical gilded bronze statues on the four pylons are by three different artists: on the right bank are *Fame and Science* and *Fame and Art* by Em-

manuel Frémiet, and on the left bank *Fame and Commerce* by Pierre Granet and *Fame and Industry* by Clément Steiner. The marble statues that stand at the base of the pylons (which are already laden with architectural decoration) are by the sculptors Gustave Michel, Alfred Lenoir, Jules Coulant and Laurant Marqueste.

Despite the criticism of 20th-century supporters of functionalism, the heavy magnificence of the Alexandre III Bridge fits very well into the grandiose city landscape of Baroque Paris between the gardens of the Champs-Elysées and Les Invalides.

64 ■ The many bronze statues at the bases of the four large pylons and at the ends of the parapets are of putti with fish.

64-65 ■ The allegorical bronze composition on the crown of the downstream side is titled *Nymphs of the Neva,* and bears the Russian coat-of-arms.

65 bottom ■ The bridge is lit up by twenty-eight lamps on the parapet and four at the entrances, and has the beautifully illuminated Invalides as a backdrop.

Location	Designed by	Length – Widest Span	Type	Construction
Paris (France)	Jean Résal and Amédée Alby	377 ft / 351 ft	Arch bridge	1897-1900

The Advent of Reinforced Concrete

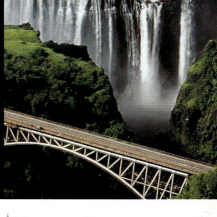

▲ Harbour Bridge   ▲ Golden Gate Bridge   ▲ Victoria Falls

▲ Golden Gate Bridge ▲ Harbour Bridge, 1930

DURING the last years of the 19th century, a new construction technique was invented that used reinforced concrete. It originated in France when, from 1849, the gardener Joseph Monier used small iron rods to strengthen the concrete from which he built flower boxes. In 1873, having perfected the method and wishing to use it in larger constructions, Monier requested and obtained a patent that also covered small arch bridges.

In 1892 another 'practical man,' the builder François Hennebique, who had begun his career as a stonecutter, patented a new method for using reinforced concrete that bettered any other. At the start of the 20th century, the use of the new technique spread across Europe: though not yet accepted in academic circles, practical people welcomed it and saw possibilities for its application.

At that time, the design of structures in reinforced concrete was guided by intuition and experience rather than calculations. Like in other branches of science, applications in the field of construction preceded the rational development of theory. During the early decades of the 20th century experiments on the behavior of the new material allowed plausible hypotheses to be developed on which calculations could be based correctly.

As in the field of large coverings, a new and profitable phase began in bridge building, in which the materials of reinforced concrete and metal rivaled one another in the construction of metal structures.

First Half of the 20th Century

Location	Designed by	Length – Widest Span	Type	Construction
Victoria Falls (Zimbabwe)	John R. Freeman	673 ft / 513 ft	Metal arch bridge	1903-1905

Victoria Falls Bridge

Zambesi River (Zimbabwe)

The Victoria Falls Bridge was part of the great project for a railroad from Cairo to Cape Town, which for political and economic reasons was never built. The bridge crosses the Zambesi just downstream of the Victoria Falls, which today are part of the Mosi-oa-Tunya park. They lie about 1,550 miles from Cape Town by railroad, on which a special *train de luxe* used to run twice a week in the early decades of the 20th century.

Cecil Rhodes, the president of the British South Africa Company, argued that the bridge should be located as close as possible to the falls to offer travelers an unmatched view combined with the excitement of plunging into the vapor cloud. The idea had its critics and was debatable for two reasons: one, because technically the point chosen was not suited to construction, and two, because a giant metal structure seemed an unacceptable intrusion into the beautiful site. In the end, however, the bridge became part of the landscape.

70 ■ In the 1930s tourists arriving at the Falls admired the smoke of the locomotive as well as the vapor of the waterfall.

70-71 ■ The wide, lovely archway, built with such effort, seems just a toy against the grandeur of Nature.

The bridge crosses a deep gorge 420 feet above the water with a span of 513 feet, and today it is also used for bungee jumping. John R. Freeman, the designer, devised a metal structure based on a reticular truss over a parabolic arc, and this was attached at the ends by means of a hinge mechanism. The foundation was made of concrete reinforced with steel bars. The height of the truss was 105 feet at the springers and 16 feet at the center point.

The chief engineer, George Camille Imbault, had to tackle huge problems associated with the difficult location of the bridge. It was necessary to build a telpher (an overhead cableway) to transport the materials between the two banks; for this, the first cable had to be launched using a rocket. Construction of the arch began from both sides, and it was necessary to add temporary fixed ends to create the attachment so as to progress with two bracket structures. For reasons of safety, an enormous net was slung beneath the work zone, but the workers found it made them nervous. It was removed as soon as possible.

Owing to a delay in the delivery of the materials, work only began in May 1904, but it proceeded quickly and on April 1, 1905 the last pieces of the arch were attached. The bridge was tested with the passing of a train weighing 612 tons at a speed of 14 miles per hour, and the center of the bridge lowered by just one inch.

Location	Designed by	Length – Widest Span	Type	Construction
Graubünden (Switzerland)	Robert Maillart	476 ft / 295 ft	Arch bridge	1930

Salginatobel Brücke

Salgina Valley, between Schiers and Scuders, Graubünden/Grisons (Switzerland)

The Salginatobel Bridge was the culmination of the art of Robert Maillart (Berne 1872–1940). Maillart was a pioneer in the use of reinforced concrete even if his entry into the bridge-building profession occurred when the new technique was still unknown. At that time two systems were in use, patented respectively by Joseph Monier and François Hennebique, both of whom were self-taught.

At that time reinforced concrete was still not studied in engineering schools, and progress in its use was the result of practical experience and instinct. Nonetheless, the improvements brought during the decades to the Hennebique method were not only due to the unquestioned genius of Hennebique himself, but also to the teams of engineers that worked directly or indirectly for his Parisian company or its numerous European branches.

Robert Maillart's career remained independent of Hennebique. He studied at the École Polytechnique in Zurich, where he graduated in 1894. After an apprenticeship at the studio of Pümpin & Herzog in Bern, he was made chief engineer for Zurich City Council, then worked for Fraté & Westermann in Zurich for two years before going solo at the age of thirty so he could work in full freedom.

Maillart's aim was to create a structural model that would satisfy static, aesthetic and budgetary demands at the same time. He was convinced that the potential of reinforced concrete was not being exploited sufficiently, and that the mistake being made by many designers was to remain tied to the traditional forms of masonry bridges without realizing that the new material offered the possibility of leaving those forms behind. It was necessary to take into account the peculiar properties of the ma-

72 ■ The arch, deck and slender vertical piers together create the monolithic structure of Robert Maillart's bridge.

72-73 ■ The bridge spans the Salgina Valley, rendering the natural harshness gentler with its elegant profile.

terial, on which he wrote: '*Reinforced concrete does not grow like wood, nor is it laminated like section iron, nor is it bound like masonry. But, as a material that is cast in molds, it might rather be compared to cast iron, whose particular forms, created from long experience, could teach us many things in the way that they pass from one construction element to another in a progressive and continuous way, avoiding sharp angles.*'

In line with his convictions, he tested new criteria when he had the chance to build two bridges in reinforced concrete in Switzerland: at Zuog in 1901 and Tavanasa in 1905. Finally Maillart had found his style. His innovation can be summarized into imperatives that were essential to the design: the bridge must be monolithic (i.e., the arch and its deck are a single unit) and the structural schema had to have great static clarity. To Maillart, the ideal model was a three-hinged arch.

At first Maillart's model did not rouse interest, but 25 years later, when the Salginatobel Brücke was built, the reception was very different: engineers and architects considered it a model to follow and began to speak of 'Maillart bridges' as a new type, and as a symbol of elegance, functionality and economy. It was 1930 and Maillart was 58 years old.

The Salginatobel Bridge lies discreetly in the landscape: it crosses a deep, wooded ravine with a span of 295 feet and a rise of 43 feet, and its slimness and lightness raise great admiration. The arch narrows at the key point to just 27 inches, then thickens at the reins (where it merges with the deck) before narrowing again at the ends in obedience to the laws of the static model. The bridge's overall lightness derives from the empty spaces created by the removal of all useless materials that would otherwise weigh upon the arch.

Harbour Bridge

Parramatta Estuary, Sydney (Australia)

Before Sydney Harbour Bridge was constructed, the north section of the city, on the left bank of the Parramatta estuary, was practically isolated from the city center. The connection between the two was made by means of a long railroad or road route that crossed five bridges.

The plan to build a crossing between Dawes Point and Wilson Point dates back to the mid-19th century, and in the following five decades 24 bridge projects and one tunnel project were proposed; but when these were examined in 1904, they were all rejected.

The problem was then studied by the Australian John Job Crew Bradfield (1867-1943), who was Chief Engineer at the Department of Public Works. He drew up a preliminary plan on the basis of which the specifications for the *International competition for an arch bridge with end piers dressed in granite* were set out, and the competition was announced in 1922. It was won by Dorman Long & Co. of Middlesbrough (England) with the design by the London engineer Sir Ralph Freeman. The construction was begun in 1926 and completed six years later under Bradfield's direction.

74-75 ■ The majestic Sydney Harbour Bridge crosses the estuary of the Parramatta River to join the north part of Sydney to the city center and modern Opera House.

Elevation of the Arch

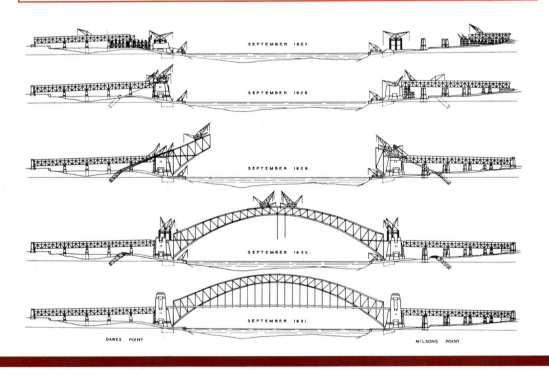

The gigantic bridge crosses the entire width of the mouth of the port and allows ships, including large liners, to pass beneath to reach the wharves inside. To emigrants arriving in Australia, the bridge was a symbol of hope, like the Statue of Liberty for those arriving in New York.

The bridge is enormous to look at: it has a single arch with a span of 1,650 feet that rises to a height of 440 feet above the water. The road deck is over 1,000 yards long and passes 160 feet above the water. In the center it is suspended from the arch, then passes between two masonry piers at the sides, each 292 feet tall.

The structure is a metal lattice, with the arch hinged at the ends, and from which a reticular tunnel truss is suspended through which the trains pass.

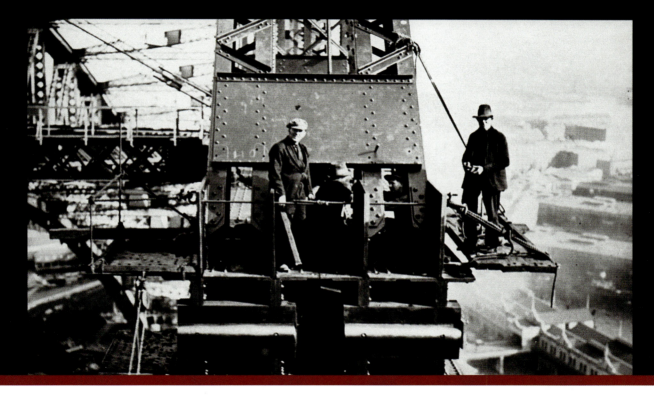

76-77 ■ Using the cantilever method, construction of the metal arch, from which the deck is suspended, was completed in 1930.

76 bottom ■ The design of the bridge was very carefully studied, not simply with regard to its details, but to establish the construction times and procedures.

77 ■ These two pictures from 1930, which show workers without helmets or insurance of any sort balancing on high scaffolding without any protection beneath, seem incredible today.

Location	Designed by	Length – Widest Span	Type	Construction
Sydney (Australia)	Sir Ralph Freeman and John Job Crew Bradfield	3,770 ft 1,650 ft	Arch bridge with suspended road deck	1926-1932

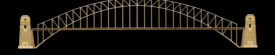

78 top ■ The unions that join the parts of the structure required more than six million rivets.

78 bottom ■ The complexity of the reticular structure is apparent to the traveler about to cross the bridge.

HARBOUR

78-79 ■ The walkways along the sides of the bridge are open to visitors who can admire the views from the top.

80-81 ■ The structural design of the bridge can be seen above the fog.

The upper level carries a highway and a pedestrian path.

Technically and organizationally, construction was very demanding. To meet the needs of the port activity, it was decided to build using the technique of cantilevering, from the springers toward the center. It was therefore necessary to create temporary fixed ends during the construction process. Testing of the bridge occurred in February 1932 with the passage of 96 steam locomotives.

HARBOUR Bridge

Golden Gate Bridge

GOLDEN GATE STRAIT, BETWEEN THE PACIFIC OCEAN AND SAN FRANCISCO BAY (USA)

The need to span the Golden Gate Strait with a bridge became apparent in 1923, but it was only as part of President Franklin D. Roosevelt's New Deal that its construction was decided, in part to boost the local economy. Between 1933 and 1937, two bridges were built in the city of San Francisco: the *San Francisco-Oakland Bay Bridge* over the bay, and the *Golden Gate Bridge* over the Strait. The poetic name 'Golden Gate' dates to 1846 and seems to have been coined by the military topographer John C. Frémont, who saw in the Pacific strait a similarity with the port of Istanbul, which was also called *Chrysoceraso*, or Golden Horn.

Construction of a bridge across the Golden Gate Strait presented sizeable technical difficulties because of the stresses placed upon the structure by the strong natural forces flowing in from the Pacific. The designer and chief engineer, Joseph B. Strauss, invited architect Irving F. Morrow to work with him on the design. The structure would have to withstand currents flowing at 115 miles per hour and wind-caused oscillations of up to 26 feet. An unplanned test occurred on December 1, 1951, when a storm blew up with winds of 80 mph. The central span registered horizontal movement of 24 feet and vertical movement of 5 feet, oscillations that the structure supported without significant damage.

82 ■ Fog is present in the Golden Gate Strait for many months each year. It thins as it rises, leaving just the top of the pylon visible against the sky.

83 ■ An aerial view of the Golden Gate Bridge reveals the heavy traffic that uses it and which makes it a vital element of San Francisco, as well as its symbol. The north–south road traffic crosses the east–west route of shipping leaving and entering the bay.

GOLDEN GATE

Also difficult was construction of the south pier's foundation at a depth of 100 feet; this required the use of a giant compressed-air caisson. Although there was a safety net beneath the deck, which saved the lives of 19 workers, 11 lives were lost during construction.

Two and a half times the length of the Brooklyn Bridge, the Golden Gate is formed by three suspended bays: the lateral two have spans of 1,066 feet, and the central one 4,200 feet. This remained the world record until 1964.

An attractive color effect is created by the vermilion-orange of the support piers, which rise 745 feet above the water level and 502 above the road deck. At the south end of the bridge and in the cen-

ter are two horns used to guide shipping in the event of fog. Between July and October, the haziest season, the horns are in use for over five hours a day. Light signals for airplanes are installed on the top of the piers.

Currently an estimated 40 million vehicles cross the bridge a year (for which a toll is paid). A permanent team of 38 painters and 17 welders carry out continuous maintenance work on the bridge.

84 and 84-85 ■ Construction of a large bridge in adverse weather conditions requires out of the ordinary building equipment.

85 bottom ■ Joseph B. Strass, the designer of the bridge's structure and the Director of Works, shows some eminent visitors (R. Clarke, C. E. Paine, C. Dungan) the important and delicate operations of positioning and tensing the cables.

Location	Designed by	Length – Widest Span	Type	Construction
San Francisco (USA)	Joseph B. Strass and Irving F. Morrow	8,980 ft 4,200 ft	Suspension bridge	1933-1937

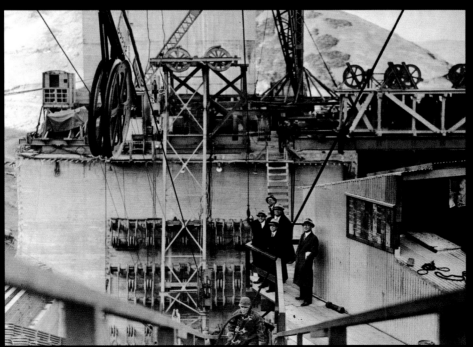

88 bottom ■ The join between the upright and crosspiece of the pylon is typical of the architectural style of the 1930s.

88-89 ■ Those who don't fear heights, can admire the view of the mountain west of the bridge from the top of the north pylon.

GOLDEN GATE

86-87 ■ A painter works protected by an elaborate harness and cable system.

86 bottom and 87 bottom ■ The reticular truss of the deck (left) can withstand the force of the wind, and the saddles (right) determine the curvature of the cables.

88 top ■ The arches are formed by a complex reticular substructure.

TOWARD the middle of the 20th century, with the invention of prestressed concrete, a turning point occurred in the construction of reinforced-concrete structures.

▲ Verrazano Bridge ▲ Puente Vasco De Gama ▲ Great Belt Link

Prestressed concrete was the realization of an idea voiced in 1888 by two German builders, Döring and Könen. It consists in compressing the concrete before use in order to compensate the forces of traction that this material is unable to resist. At that time it was not possible to act on the idea as the means available could only produce a very slightly compressed product that would have quickly been negated by the effects of shrinkage and creep. The principle was taken up again thirty years later by Eugène Freyssinet. From 1928, the year in which he applied for his first patents, Freyssinet devoted himself to improving his invention, using his brilliance and intuition to fill in the gaps where science was lacking; however, these gaps in knowledge caused spitefulness and backbiting from some of his less objective competitors.

During the 1940s the use of prestressed concrete began to spread slowly, and the new material received official recognition in 1953 with the establishment of the International Federation of Prestressing, backed by the most authoritative names in civil engineering in Europe. Research into the behavior of concretes and steels intensified during this period as greater knowledge enabled better usage, making possible better designs, in particular of bridges.

Over the last fifty years there has been an almost frenetic sequence of increasingly demanding structures. Older types, like suspension and cable-stayed bridges, have returned, having been given a new lease of life by more advanced technologies, and they continue to increase in size. Around the world huge bridges are being built to connect islands, states and even continents.

Aesthetically, the most recent bridges have been spectacular because of their very size and the drama they add to the landscape. The structures are becoming increasingly pared down, the best examples adapting to the morphology of the land, and combining elegant curves with straight lines. In other cases the designers restrict themselves to schematic, rigidly orthogonal designs that are not associated with the bridges' contexts.

From 1950 until Today

Location	Designed by	Length – Widest Span	Type	Construction
Maracaibo (Venezuela)	Riccardo Morandi	28,543 ft 771 ft	Cable-stayed bridge	1958-1962

Puente Maracaibo or Puente General Urdaneta

Maracaibo Lake, Zulia (Venezuela)

The General Urdaneta Bridge was built between 1958 and 1962 during the government led by Rómulo Betancourt, which was elected after the downfall of the dictator Marcos Pérez Jiménez. During the same period, the road network was expanded and two other bridges were built: the General Páez Bridge over the Apure River and Angostura Bridge over the Orinoco on the road to Guyana. These latter two bridges were large projects but of less importance than the long bridge that crosses the strait and connects Maracaibo with the rest of Venezuela.

The Maracaibo Bridge, which is 5.4 miles long, was designed by Riccardo Morandi (1902–89), a famous name among Italian engineers. Shortly after World War II, Morandi built a number of bridges in Italy using prestressed concrete, contributing greatly to the material's improvement and development. Morandi's theoretical studies (deriving from his work as a professor), combined with his professional experience as a builder, led him to pin down a static model that typified his style. Exploiting all the technical possibilities, he was able to create structures of high engineering quality and aesthetic appearance that brought him fame and important commissions, of which the Maracaibo Bridge is one.

92 ■ On December 21, 2002, the oil tanker *Pilin Leon*, about to pass beneath the bridge without a permit, was rerouted to the port of Bajo Grande and handed over to a Civil Court.

92-93 ■ The pylons of the five cable-stayed spans across the navigable channel of the lake break the monotony of the long bridge.

94-95 ■ The piers of the two lateral viaducts gradually rise to the height of the bridge in the navigable section.

Morandi's design for the bridge over the Venezuelan lake differed from the other proposals submitted to the government in that it was made from reinforced concrete rather than steel. This choice was less costly from the standpoints of construction and the greater resistance to the tropical climate.

In total the bridge has 135 spans, with two viaducts made from trusses that connect the two shores to the middle section of the bridge. The middle section is formed by 5 cable-stayed spans of 771 feet each that cross the navigable part of the strait. The entire structure, including the stays, is made from prestressed concrete. The piers, like the supports of the trusses at either end of the bridge, are 302 feet high and they continue the technical design of the trusses formed by a combination of variously shaped triangles.

Morandi's designs are famous for their elegant looks but, in the case of the Maracaibo Bridge, the use of a modular structure meant that standard elements – either prefabricated or produced on site – could be used, which also brought economic benefits.

The Maracaibo Bridge was damaged in April 1964 when an Esso supertanker ran into it, destroying piers 31 and 32 and cracking numbers 30 and 33. In 1999 an inspection revealed cracks in other piers, but of minor extent.

PUENTE MARACAIBO

Verrazano Narrows Bridge

SPANNING THE VERRAZANO NARROWS NEW YORK (USA)

The idea of building a bridge over the Verrazano Narrows dates back to the early 20th century. Initially a tunnel was proposed, and this idea was warmly supported by the mayor of New York, John F. Hylan. As soon as government approval was received, Hylan initiated excavation operations. However, costs turned out to be much higher than expected and digging was halted. Today the abandoned entrances to the tunnel remain and are still used as an ironic allusion to Mayor Hylan. In 1926 a design for a suspension bridge presented by the well-known engineer David Steinman was examined, but financing was refused.

After more than another thirty years, the need for a new bridge was recognized; it was eventually built by Ammann & Whitney between 1959 and 1964. The design was by Othmar Herman Ammann, who concluded his brilliant career with this bridge. Ammann was born on March 26, 1879 in Feuerthalen, Switzerland, and he graduated from Zurich Polytechnic where he had studied under Wilhelm Ritter. Two years later Ammann emigrated to the United States. There he was so successful as an architect that in 1932 New York University awarded him an *honoris causa* degree, and in 1937 the Penn-

96 ■ The New York Marathon crosses the Verrazano Bridge that links Brooklyn to State Island and is part of the 'bypass' that avoids Manhattan. An average of 190,000 vehicles use it every day.

97 ■ The bridge is supported by four cables 36 inches in diameter, each formed by 26,108 strands. Vehicles cross on two six-lane levels.

98-99 ■ The two steel pylons weigh 26,000 tons each and stand 680 feet above the water. Due to the curvature of the Earth, the tops of the pylons are fifteen hundredths of an inch farther apart than the bases.

98 bottom and 99 top ■ In 1962 the complex and dangerous phase of stringing out the cables was carried out. The procedure was made easier by the construction of a special pair of temporary walkways, but nonetheless three men lost their lives in accidents.

sylvania Military Academy did the same. Among the many great bridges Ammann built are the famous Golden Gate and George Washington bridges. He died in New York on September 22, 1965.

The Verrazano Bridge crosses the Narrows with an overall length of 4,567 yards, and a structure 5,250 feet long. This huge bridge is made almost entirely from steel and was awarded first prize by the American Institute of Steel Construction in 1965. The section that crosses the water has three spans suspended from cables anchored in the banks, and supported by two piers that stand in the water. The central span of 4,258 feet held the record for width until 1981. The bridge carries a highway that, is 108 feet in width and is on two levels, each of which carries six traffic lanes.

The bridge was named after the Italian explorer Giovanni Verrazano (1485-1528), who was the first European navigator to enter New York Bay.

Location	Designed by	Length – Widest Span	Type	Construction
Verrazano Narrows Bridge (USA)	Othmar Hermann Ammann	13,700 ft 4,259 ft	Suspension bridge	1959-1964

Location	Designed by	Length - Widest Span	Type	Construction
Kingston-on-Hull (Great Britain)	Gilbert Roberts and Bill Harvey	7,283 ft 4,626 ft	Suspension bridge	1972-1981

Humber Bridge

Spanning the Humber Estuary at Kingston-on-Hull (Great Britain)

By the mid-19th century the need to join the two sides of the Humber estuary had become clear. The first real plan put forward in 1872 and called for a tunnel. Later, various bridge designs were considered but when, in 1928, a plan was finally placed on the table, the bridge could not be built owing to the grave economic situation in Europe and the United States.

The possibility of constructing the long-awaited bridge came in 1959 when the Humber Bridge Act set up a commission to build the structure. The alternative of a tunnel had been excluded because the geology of the subsoil made the cost of construction excessive. The commission decided on a wide suspension bridge that would not obstruct the river traffic.

Designed by Gilbert Roberts and Bill Harvey, the Humber Bridge is 2,428 yards long. The suspended span is 4,626 feet in length and flanked by two viaducts that extend 306 yards to the north and 580 to the south. Each of the support pylons is made of reinforced concrete and formed by columns 508 feet high connected by four

crosspieces 72 feet long. The deck rests on the lowest and thickest crosspiece so that the road level is 98 feet above the water; the highest crosspiece is 26 feet from the top of the pier.

The deck is made of metal, with a hollow pseudo-trapezoidal section and two brackets. The bicycle and pedestrian lanes are each 10 feet wide and pass on the outside of the piers. The four-lane highway is 72 feet wide. Construction began in 1972 and lasted 8 years.

The bridge, for which the local population had waited for more than a century, eventually opened on July 17, 1981; each year more than six million vehicles cross it.

100 ■ The bridge's main span allows the passage of river traffic. It stretches between the pylon in the water and the north pylon on the left bank of the estuary, close to the city of Hull.

101 ■ The north and south (seen here) ends of the access viaducts to the bridge rest on the massive counterweights that anchor the suspension cables.

Puente Lusitania

Guardiana River, Mérida (Spain)

The Lusitania Bridge joins the old part of the city of Mérida in Extremadura to the districts north of the River Guardiana. The designer was the artist, engineer and architect Santiago Calatrava. It was necessary to find an individual with the awareness and culture of Calatrava to build a bridge in the ancient *Emerita Augusta,* the Iberian city with the most Roman monuments from the Augustan era: Tiberius' Arch, the theater, the tombs of the Voconii and the Julii, the temples of Diana and Mars, three aqueducts and two bridges. The latter cross the Alboregas and Guardiana rivers.

Santiago Calatrava was born in Valencia, Spain, on July 28, 1951. He has an impressive record: after studying art, he graduated in architecture and, attracted by the precision of the mathematics of the greatest architectural works, he decided to round out his scientific knowledge at the Zurich Polytechnic, where he graduated in civil engineering in 1979. He soon achieved international fame, and has designed more than 40 large structures around the world, many of which are bridges.

102-103 ■ The partial illumination – designed to highlight the bridge's design – guides the observer's eye to the most important details.

102 bottom ■ The steel and concrete arch that supports the structure embraces the central section of the bridge used as a pedestrian walkway.

103 top ■ The refinement of the detail in the design of the Lusitania Bridge means that structural elements like the stays can seem like elegant features to the untrained eye.

103 bottom ■ The new, almost transparent bridge discreetly inserts itself between ancient Roman structures and demonstrates the progress made in bridge technology.

The Lusitania Bridge was built between 1988 and 1991. It has two straight lateral sections and a central span with a low arch. Measuring 620 feet in span, the arch lies over the pedestrian way that divides the two carriageways. The arch is a combination of a steel latticework and two sections made of reinforced concrete at the springers. The ends of these concrete sections (which terminate in a horizontal section set in the deck truss) each have a portal through which the pedestrian path passes. The deck is 508 yards long overall and formed by a reinforced concrete truss. This is suspended from the arch in the central span, while in the lateral sections it rests on the truss. The steel ties are inclined and the bearings are formed by short dual elements with variable circular sections.

The Lusitania Bridge is simple and elegant. It is a rare example of a structure characterized by lightness though not without a sense of monumentality. It seems worthy to lie alongside the ancient Roman bridge, which is less than a mile upstream.

Location	Designed by	Length – Widest Span	Type	Construction
Mérida (Spain)	Santiago Calatrava	1,526 ft 620 ft	Arch bridge with suspended deck	1988-1991

104-105 ■ La Barqueta footbridge provides the principal entry point to the Isla Mágica theme park and Seville Technology Center on La Cartuja island.

104 bottom ■ The depressed arch is supported at either end by pylons inclined to increase the width of the bridge and which also act as portals.

105 right ■ The form of the connection joints between the pairs of pylons and the arch represents the perfect fusion of geometry and mechanics. The position of the joints means that they are the structure's fundamental points.

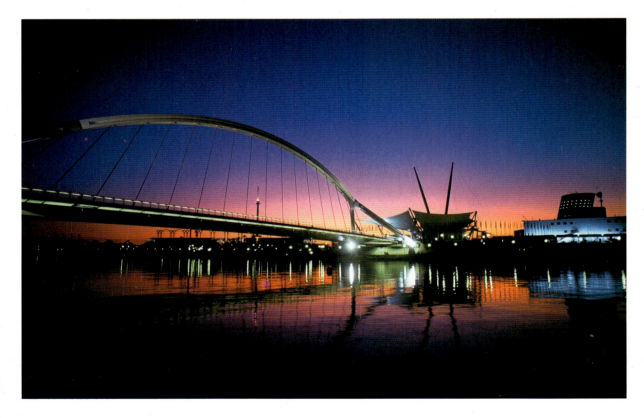

Puente la Barqueta

Guadalquivir River, Seville (Spain)

The Barqueta Bridge connects the historical center of Seville to the island of La Cartuja. It was built between 1989 and 1992 to give access to the theme park *Isla Mágica* for Expo92, the world exhibition that celebrated the 500th anniversary of the discovery of America.

The design for the bridge was by Juan Arenas de Pablo and Marcos Pantaleón, who in the last fifteen years have shown themselves to be some of the world's most aesthetically and environmentally aware engineers. According to Arenas de Pablo, an engineer is not only responsible for the functionality of his works but must also give his best in terms of aesthetics: *'The [designer's] mental organization of the structure is a conceptual operation parallel to the ideation of the external appearance of the construction. Not in the details, of course, but in the general conception.'* In other words, the aesthetic value of a bridge or any engineering work is not simply something added to the structure, but part of the whole by which structural architecture is judged, and this, by definition, combines *venustas* (beauty) and *firmitas* (stability). In practice, Arenas sticks to his convictions, and La Barqueta Bridge is a fine example in the way that *venustas* and *firmitas* are perfectly integrated.

The load-bearing structure is made completely of metal, and is formed by a low arch (aligned with the center line of the deck) that rises from two inclined 'legs' at either end. These form a triangle, the base of which is formed by the deck, and the vertex the point from which the arch begins. Structurally, the piers reduce the span of the arch from 551 feet (the entire span) to 354 feet. Aesthetically, the triangles appear as large, inviting gateways to the bridge. All the elements have hollow rectangular sections, with moldings that have the dual purpose of stiff-

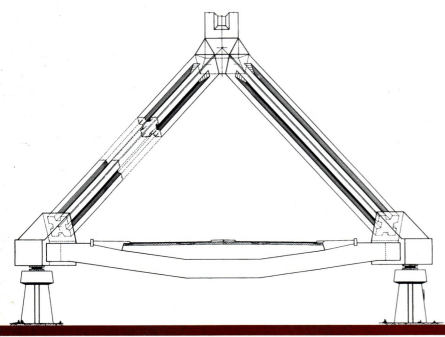

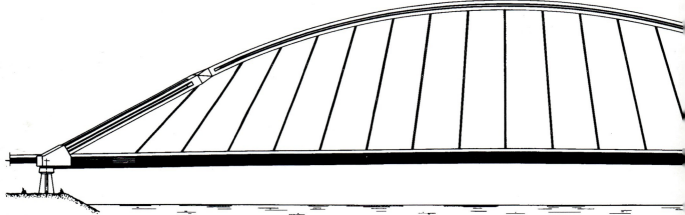

ening the structure and creating visually stimulating plays of light and shadow.

The deck is supported by suspension ties attached to the center line of the deck that also act as a dividing line between the two carriageways. The fact that the arch is shorter than the length of the deck means that the ties converge as they rise, and this creates a sense of dynamism that matches the movement of those crossing the bridge.

106 and 106-107 bottom ■ The sketches show the profile of the arch, the direction of the suspension cables and the static arrangement of the system – formed by the inclined pylons, the cross-piece and the rests – that connects the arch to the ground.

106-107 top and 107 bottom ■ The square steel elements have deep grooves that serve to provide longitudinal reinforcement. The converging red stays create a sense of dynamism.

Location	Designed by	Length – Central Span	Type	Construction
Seville (Spain)	Juan J. Arenas de Pablo and Marcos J. Pantaleón	551 ft / 354 ft	Metal arch bridge with suspended deck	1989 / 1992

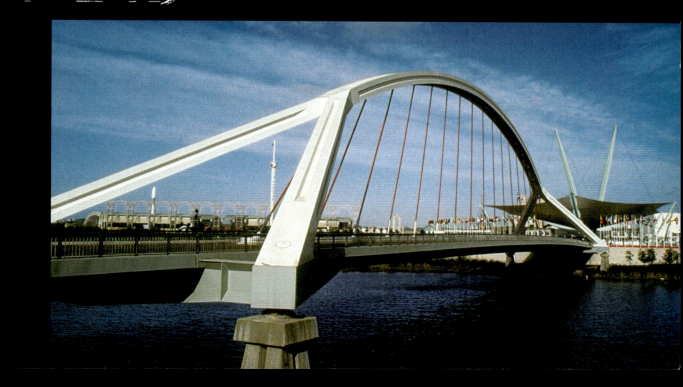

Pont de Normandie

River Seine, Le Havre (France)

Normandy Bridge crosses the Seine estuary between Le Havre and Honfleur. Besides connecting the two sides of the river — the right heavily industrialized and the left mostly touristic — the bridge reduces the distance between Lower Normandy and Le Havre by 25 miles and connects with the estuary highway. It makes a decisive contribution to the region by helping to integrate the local area in the national and European economies.

The importance of the bridge and its technical complexity prompted the Chamber of Commerce in Le Havre (the body responsible for the city's bridges and which commissioned the new struc-

108-109 ■ The thin ribbon that gets lost is a sturdy steel sheet structure. The transversal section of the trapezoidal deck is more than 69 feet wide and 10 feet high. These are notable dimensions but they are negligible compared to the bridge's length.

Location	Designed by	Length – Widest Span	Type	Construction
Le Havre (France)	Michel Virlogeux, François Doyelle, Charles Lavigne, Brice Girard	7,024 ft / 2,808 ft	Cable-stayed bridge with mixed structure	1992-1995

110 top ■ Four stays on each anchorage level are joined to the pylon: two support the central truss over the river and two are anchored to the deck of the viaduct.

PONT DE NORMANDIE

110 bottom ■ The stays are connected to the supporting metal structure inside the pylon, hidden in a concrete shell.

111 ■ The road deck passes between the two legs of the pylons and rests on the crosspieces 135 feet above the base.

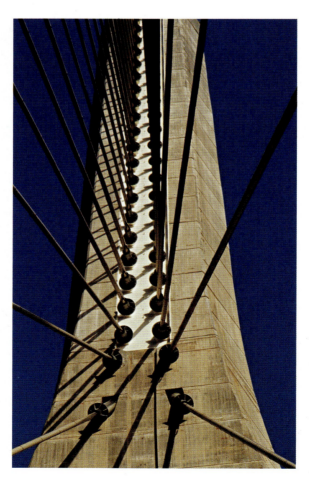

ture) to turn to the departmental *Équipement de Seine-Maritime*, which made use of SETRA *(Service d'études techniques des routes et autoroutes)* and the private sector. The preliminary studies and plans were carried out in 1989, and these were then submitted to an international scientific panel. In 1990 excavation began on the foundations and the entire bridge was built between 1992 and 1995.

Normandy Bridge held the record for the longest cable-stayed bridge until 1999, when the record was taken by the Tatara Bridge, Japan. The Normandy structure is 1.3 miles long in total and has a central span of 2,808 feet. There are also two lateral viaducts, the one to the south 599 yards long, and the one to the north 807.

The viaducts are composed of continuous caisson trusses in prestressed reinforced concrete that

112-113 ■ The most delicate moment during assembly is the hoisting of a new piece with the cranes and its attachment to the deck.

113 top right ■ To reduce the effect of the wind, the stays are lined with polyethylene sheaths to diminish the friction coefficient.

113 bottom ■ The first construction stage was achieved in spring 1993 with the completion of the two 705-foot pylons.

continues for a hundred yards or so into the central span. This is cable-stayed with a deck of mixed construction: the middle section in steel caissons is 2,047 feet long, and lies between sections of prestressed concrete that are connected to the viaducts.

Overall the piers are 705 feet high and have a mixed structure in which the metal core is lined by two half-shells in reinforced concrete. The lower section is in the shape of an elongated letter A with its tip at a height of 459 feet. Above this is the vertical top to which the cable-stays are attached.

During the three years of construction, the site proved to be an attraction to the public as well as to engineering students and professionals. To answer all the queries that were put to it, the Le Havre Chamber of Commerce and the *Mission Pont de Normandie* decided to employ specialized staff and allow the public to tour the site on special days (the *Journées chantier ouvert*): the scheme was amazingly successful and created a new phenomenon: 'technical tourism.' It attracted an unpredictable number of visitors that averaged out at around 70,000 a year.

PONT DE NORMANDIE

114-115 Ninety-two radiating stays connect the deck of the central span to the two support pylons.

115 bottom The lighting at night shows up the contrast between the apparent lightness of the deck and the solidity of the pylons.

■ OHNARUTO BRIDGE   ■ AKASHI KAIKYO BRIDGE

■ HITSUISHJIMA-SETO BRIDGE   ■ IWAKUROJIMA BRIDGE

■ KURUSHIMA KAIKYO BRIDGE   ■ OMISHIMA BRIDGE

THE BRIDGES OF THE THREE HONSHU-SHIKOKU CONNECTIONS

(JAPAN)

In 1975 a project was undertaken to connect Honshu, Japan's largest island, with Shikoku, another of the four main islands. The Honshu-Shikoku Bridge Authority built the connection between 1976 and 1999.

The project was based on three large roads that more or less run north-south: the Kobe-Naruto route, the Kojima-Sakaide route, and the Honomichi-Imabari route, which are positioned respectively close to 135°, 134°, and 133° southeast of the Greenwich meridian.

The project features 18 bridges, almost all made with large-span structures.

The most prevalent types of bridge are suspension (10) and cable-stayed (5), and both types include the current record-holders in terms of span width. The three highways connect the many smaller islands in the Seto Sea and follow the routes created by the position of the islands.

The first suspension bridge, the Innoshima, was built between 1977 and 1983. Its successful outcome led to the adoption of the same criteria for the designs of all the suspension bridges.

■ The Kobe-Naruto route is a highway 55 miles long built between 1976 and 1998. It passes through Awaji island and crosses two very wide straits: the Akashi (2.5 miles) and Naruto (1,420 yards). There are only two bridges on this route: the Akashi Kaikyo Bridge and Ohnaruto Bridge.

■ The Kojima-Sakaide route is 24 miles long and also known as the *Seto-Chuo Expressway & JR Seto-Ohashi Line* as the highway and railroad run above one another for 20 miles. The road crosses six straits and passes through five

■ KITA BISAN-SETO BRIDGE ■ SHIMOTSUI-SETO BRIDGE

■ OHSHIMA BRIDGE ■ TATARA BRIDGE ■ SHIN ONOMICHI BRIDGE

117 bottom ■ To build the three connections between Honshu (Japan's largest island, on the left of the illustration) and Shikoku, eighteen bridges were required.

small islands that lie almost in a line. The set of six bridges is unique and is called the Seto-Ohashi Bridge. The individual sections are different in type and size. In order from north to south they are:

■ The Shimotsui-Seto Bridge, Hitsuishjima-Seto Bridge, Iwakurojima-Seto Bridge, Yoshima Bridge, Kita Bisan-Seto Bridge and Minami Bisan-Seto Bridge.

■ The Honomichi-Imabari route is the westernmost of the three. It crosses eight islands in the Seto Sea, several of which are much bigger than those on the Kojima-Sakaide route.

The 37-mile road passes over 10 bridges of different lengths and types: the Shin Onomichi Bridge, Innoshima Bridge, Ikuchi Bridge, Tatara Bridge, Shin Onomichi Bridge, Omishima Bridge, Hakata Oshima Bridge, Oshima Bridge and Kurushima Kaikyo Bridge. The last represents three suspension bridges in a series over the Kurushima Strait.

The most important bridges in the entire project are the Akashi Kaikyo, which is currently the bridge with the widest span in the world, and the Tatara, which has the widest span of any cable-stayed bridge.

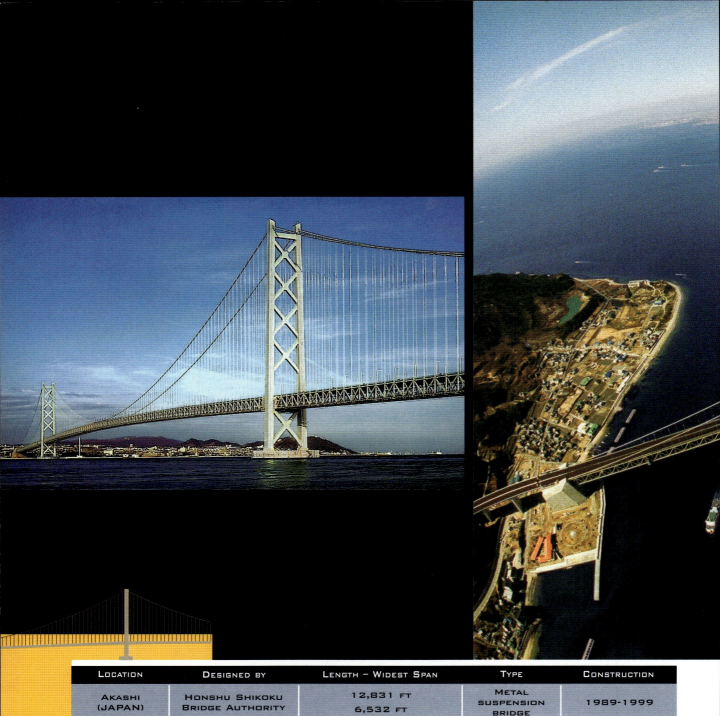

Location	Designed by	Length – Widest Span	Type	Construction
Akashi (Japan)	Honshu Shikoku Bridge Authority	12,831 ft 6,532 ft	Metal suspension bridge	1989-1999

Akashi Kaikyo Bridge

Akashi Strait (Japan)

The Akashi Kaikyo (4,277 yards) is the longer of the two bridges on the Kobe-Naruto route and connects the islands of Honshu and Awaji. This marvelous suspension bridge has three spans and is the record holder for the longest span in the world. When completed, it was longer than the Humber Bridge in England (the previous holder) and only just longer than the Great Belt East in Denmark, which was constructed at the same time as the Akashi Kaikyo. The lengths of the largest span of each of the three bridges are respectively 6,532, 4,626, and 5,328 feet.

The construction type was chosen to suit the heavy marine traffic of the mile-wide channel and the movements of the fishing boats on both shores.

The enormous dimensions of the bridge over the Akashi Strait and the environmental conditions of the location meant that extreme safety margins had to be built in. The designers had to allow for the strong winds and frequent and considerable seismic activity to which the structure would be exposed. The bridge was an opportunity to perfect new anti-seismic calculation methods and to test new ideas whose aim was to eliminate the torsional vibrations caused by the wind. 'Stabilizers' were inserted in the deck that directed the wind in order to balance out the asymmetrical pressures caused by wind over surfaces, and additional masses were inserted in the metal piers (928 feet high) to reduce the vibration in the structure during typhoons. The bridge's aerodynamic stability was tested experimentally in a wind tunnel on a three-dimensional model on a 1:100 scale.

118 ■ Having to deal with exceptionally difficult conditions, the designers of Akashi Bridge succeeded in resolving the technical problems without neglecting the bridge's aesthetic appearance.

118-119 ■ The deck is hung from the suspension cables and rests at either end on anchoring blocks. It is made up of a reticular truss that resembles the structure of the piers.

In addition to the design difficulties, there were the inevitable problems of execution in a project of this size. The geologically unsuitable terrain and strong currents in the strait created accidents during construction of the foundations of the piers, which was carried out using metal caissons 240 feet in diameter. The wind was a constant problem during installation of every part of the bridge, particularly the tall piers.

Construction, which began in 1989, lasted ten years, and the structure was obliged to withstand an unexpected test: on January 17, 1995, a strong earthquake with its epicenter between the two piers of the bridge had the effect of lengthening the distance between the piers by three feet, though it did not damage the parts already built. The design was revised in light of the new situation, and the project completed without harm coming to any of the workers. On April 5, 1998 the Akashi Kaikyo Bridge was opened to traffic.

Built completely in steel, the structure comprises two support piers, suspension cables and the deck. Each pier is formed by two elements that converge slightly as they rise, and connected by 5 diagonal crosses and 2 horizontal crosspieces. The cables could be relatively thin because of the high resistance of the steel used. The reticular truss of the deck also seems slim when compared to the length of the bridge and height of the piers.

The gray-green color was chosen to match the bridge to the urban landscape and provide a clear contrast with the intense colors of the sea and sky.

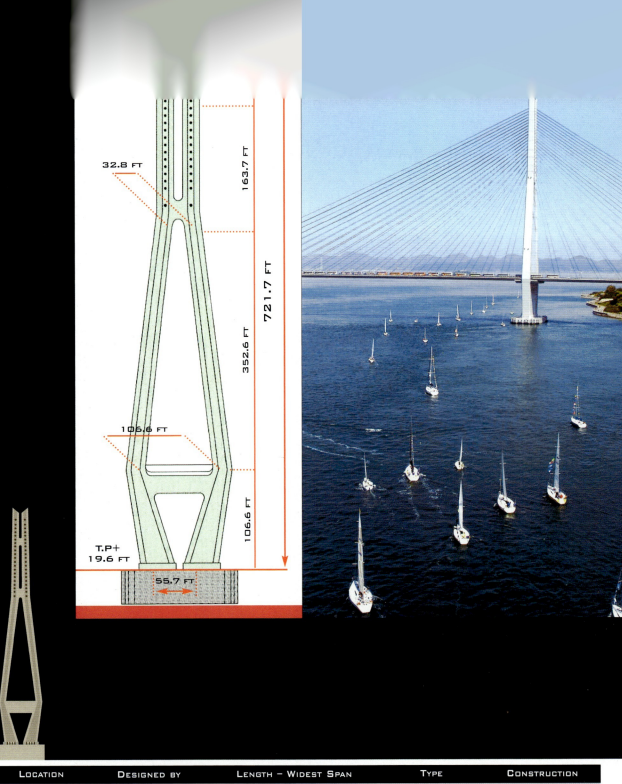

LOCATION	DESIGNED BY	LENGTH – WIDEST SPAN	TYPE	CONSTRUCTION
SETO SEA (JAPAN)	HONSHU–SHIKOKU BRIDGE AUTHORITY	4,856 FT 2,920 FT	CABLE-STAYED BRIDGE	1990–1999

TATARA BRIDGE

SETO SEA (JAPAN)

The Tatara Bridge connects the islands of Ikuchijima and Ohimishima and was built between 1990 and 1999. It is the most notable of the ten bridges on the Honomichi-Imabari route owing to its length (1,619 yards) and because it is the cable-stayed bridge with the longest span in the world (2,920 feet). This exceeds the Normandy Bridge in France by 112 feet, which held the record until 1995.

The bridge has three spans: the two end ones (886 and 1,050 feet) are made from heavy, prestressed reinforced-concrete girders that provide a counterweight to the central steel span of 2,920 feet. This span has a hollow cellular section with two lateral cantilevers: the four-lane highway runs on top of the central section (72 feet wide by 8 feet and 10 inches high), while the cycle and pedestrian lane lies on the cantilevered sides (14 feet and 5 inches wide).

The reinforced-concrete piers rise 741 feet above the water; the section below the deck is V-shaped while the upper sec-

120 ■ The design of the pylon of Tatara Bridge took into account aerodynamic demands and is the most harmonious of the scientifically acceptable forms.

120-121 ■ The plain, clean lines of the world's longest cable-stayed bridge are seen over the Seto Sea in the melancholy light of sunset.

tion is like an upturned Y. The leg of the Y is split into two parallel parts that end obliquely.

The successful combination of technical and aesthetic requisites is the result of studies by structural engineers, architects, and artists.

After the decision was taken to use the form of an upturned Y, four designs were produced: The oblique cut was included for aerodynamic reasons and appeared the best solution from an aesthetic viewpoint. To avoid the impression of discontinuity due to the density of the stays in the three spans, a transition zone was created that mitigates the effect.

Construction of the Tatara Bridge was complicated by all the problems that a project of this size brings, particularly as it is located in an area of strong winds and frequent earthquakes.

During the nine years of construction, several elements of the design were modified owing to the results of experimental trials conducted in a wind tunnel on a model scale of 1:50.

122-123, 122 bottom ■ The Tsing Ma suspension bridge, the Ma Wan land viaduct, and the cable-stayed Kap Shui Mun bridge together form part of the Landau Link that connects Hong Kong with its airport.

123 right ■ The Tsing Ma suspension bridge is the most important structure in the Landau Link and is the longest bridge in the world to carry both road and rail traffic.

Tsing Ma Bridge

Ma Wan Channel, Hong Kong (China)

The Tsing Ma Bridge joins Tsing Yi and Ma Wan islands and is part of the Lantau Link that connects Kowloon and Lantau Islands. The link also comprises the Ma Wan viaduct (an overpass 230 feet high that crosses the island) and the Kap Shui Mun Bridge between Ma Wan and Lantau.

The Tsing Ma Bridge is the most important section of the connection. It was built between 1992 and 1997 from the design by Yee Associates and is the longest bridge that carries both road and rail traffic.

The construction type chosen was that of a suspension bridge. It has a central suspended span of 4,518 feet and two lateral spans, one of which is suspended and one supported from beneath. The suspension cables are made from high-resistance steel and have a diameter of 39 inches. They are anchored at either end in massive reinforced-concrete structures and are supported by piers 676 feet tall. Neither of the piers stands in the water: the one on the Tsing Yi side is on the island itself, and the other was built on an artificial island about 130 yards from Ma Wan island.

The reinforced-concrete piers are formed by two cylindrical columns connected by four horizontal crosspieces.

Location	Designed by	Length – Widest Span	Type	Construction
Hong Kong (China)	Mott MacDonald and Yee Associates	7,218 ft / 4,518 ft	Suspension bridge	1992-1997

124-125 ■ The two suspension cables (diameter 3 feet 7 inches) each comprise 33,000 galvanized steel strands one fifth of an inch thick.

The top of each column has a cast-iron saddle in which the suspension cable is inserted; the various sections of the steel deck were prefabricated.

The components of the bridge were made partly in Great Britain and partly in Japan, then assembled in Dongguan in China. There are 96 modules, each 59 feet high, 180 feet wide and 26 feet tall. The section is a hollow rectangle 135 feet wide to which two triangles are attached at the sides.

The six-lane highway runs on the outside of the tunnel,

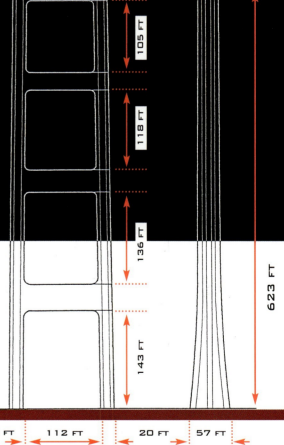

while inside there are two railroad tracks at the center. On either side of the tracks are two one-way lanes used to provide access to the tracks for maintenance purposes or as emergency lanes for road traffic. The electrical cables and an inspection corridor run alongside in side ducts.

The installation of the deck was the last phase of construction. The modules were transported in pairs on a flatboat, then towed to the right position and raised.

They were then connected to the ties that hung from the previously positioned suspension cables.

By night the Tsing Ma Bridge – the symbol of Hong Kong – is lit up with colored lights.

124 bottom ■ The two suspension cables (diameter 3 feet 7 inches) each comprise 33,000 galvanized steel strands.

125 top ■ The pairs of reinforced-concrete pylons are connected by high transversal strengthening girders.

125 bottom ■ The pylons rest on shallow rock and reach a height of 676 feet above the water level.

Great Belt Link

Between Funen (Fyn) and Zealand (Sjælland) Islands (Denmark)

126-127 top ■ The pylons of the bridge are 833 feet in height. From the top, one can see Sprogo Island to the west and, further away, at the end of the west bridge, Funen Island.

126-127 bottom ■ The Great Belt Link is in two parts, interrupted by Sprogo Island. The east section has a suspension bridge for road traffic and a tunnel for trains.

The Great Belt Link is one of the most important engineering works undertaken in the 20th century, both socioeconomically and technologically. The Link is formed by two prestressed, reinforced-concrete viaducts – one for rail traffic, the other for road traffic – that run from Funen Island to Sprogo Island, a distance of roughly 4 miles. This section is the west part of the link and is referred to as West Bridge. Continuing toward Zealand Island, the train enters an underwater tunnel while the road viaduct enters the East Bridge, formed by a suspension bridge with a viaduct at either end.

The Great Belt Link was built between 1991 and 1998 and was conceived at the same time as the Øresund Bridge between Denmark and Sweden (1995-2000) to complete the missing section of the ring between Norway, Sweden, Denmark and Germany. This link had been awaited with impatience by Scandinavian industry, which

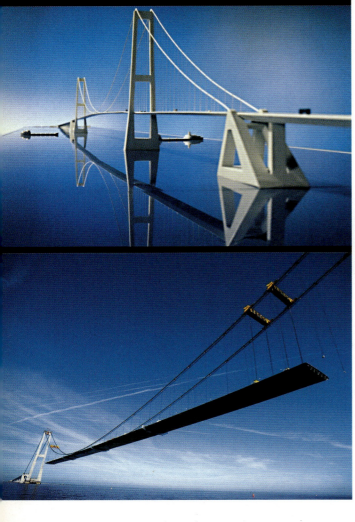

128 top left ■ Using a model, Dissing+Weitling was able to appraise the aesthetic impact of the parts of the bridge between the anchorages

128 bottom left ■ The platform for the spans was built beginning from a central position: the individual parts were raised and hooked into place at either end symmetrically.

128 right ■ The method adopted by Dissing+Weitling to study the aesthetic aspect of the design was to analyze the effects of light on sketches of the structure.

129 ■ The strands of the cables are separated and fixed individually close to the anchorages.

130-131 ■ The cables are arranged on the top of the pylons in front of the stretch of sea that separates Funen and Zealand islands.

131 bottom ■ Night lighting creates a spectacular effect on the bridge that stands in the middle of the sea.

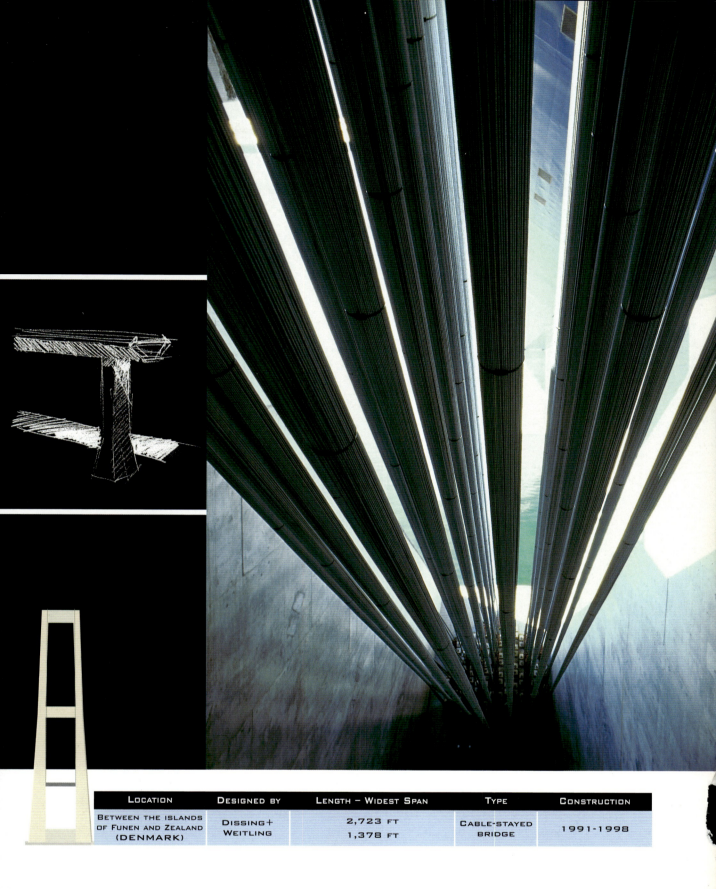

Location	Designed by	Length – Widest Span	Type	Construction
Between the islands of Funen and Zealand (Denmark)	Dissing+ Weitling	2,723 ft / 1,378 ft	Cable-stayed bridge	1991-1998

thought that trade between the south coast of Spain and west coast of Norway would be boosted.

The two viaducts of the East Bridge (to the east and west of the suspension bridge) are respectively 8,300 and 5,046 feet long. The most important section of the link is the suspension bridge with its central span of 5,328 feet (exceeded only by the Akashi Kaiko Bridge in Japan at 6,529 feet) and two lateral spans of 1,755 feet. The pylons are made of reinforced concrete and the deck from steel. The pylons rise 833 feet above sea level and are formed by two hollow tapering columns connected by two transversal pieces. The alteration in rectangular section of the columns is hardly visible in the upper part, but is more pronounced beneath the road level. This manages to combine an aesthetic appearance with the technical requirement of resisting the loads created from the deck by the force of the wind. Special attention was paid to the looks of the suspension cable anchorages: to avoid the slenderness of the bridge contrasting with their solidity, the blocks were divided into separate triangular sections.

Øresund Bridge

Øresund Strait, between Denmark and Sweden

The idea of building a connection between Denmark and Sweden dates right back to 1888 when a tunnel beneath the sea was discussed, though this, like later proposals, was never acted upon. One of the projects was blocked by World War II and taken up again in the 1960s, but it never came to anything.

In 1991 the two governments reached an agreement on a permanent connection over the Øresund. The treaty specified that the criteria for the work must be 'ecologically sustainable, technically possible and financially reasonable so as to avoid any harmful impact on the environment.'

132 ■ The cable-stayed bridge has two pairs of pylons that rise approximately 660 feet without any transversal connection above the road level. The cables are set out in parallel.

133 ■ The Øresund fixed link between Copenhagen and Malmo is 10 miles long from coast to coast. In the first fifteen months after it was opened, the bridge was crossed by more than 5 million people by car and train.

The Øresundskonsortiet (consortium) came into being in 1992; it was made responsible for raising the finance and organizing the design, construction and operation of the bridge between Copenhagen and Malmo. The same year, the consortium announced an international competition for the design, which was won by a group of societies directed by Øve Arup and Partners.

The viaduct follows a wide C-shaped curve 1,194 yards long, and has the Øresund Bridge at the center. This crosses the Filtrannan Channel in a single span of 1,607 feet. The uniform line of horizontal girders beneath the deck gives the bridge a sense of unity. The reinforced-concrete plinths that support the deck truss blend well with the anchoring piers (656 feet high) of the cables.

Each section of the deck truss is a simple, sturdy, prefabricated object; each is constructed to form a reticular

Location	Designed by	Length – Widest Span	Type	Construction
Øresund Strait (Denmark–Sweden)	Georg Rotne	3,583 ft / 1,608 ft	Metal bridge with tubular trusses	1995-2000

134 top ■ During construction of the pylons, the building of the platform also progressed, supported by temporary piers.

134 bottom ■ The long elegant viaduct and high stays visible from a long distance are a surprise to travelers arriving by road.

structure with a section in the shape of an upturned trapezoid that is sufficiently resistant to be the deck of a cable-stayed bridge. The high-speed train runs on twin tracks in the lower central section of the trapezoid (a sort of tunnel) while the four-lane highway runs above it.

The cables are arranged in the shape of a harp and are made more visible by the verticality of the stays. The effect is accentuated by the fact that each of the cable planes is supported by one of the two independent piers that are connected to one another only beneath the deck.

A last construction detail that increases the similarity to a harp is the direction of the stays, which coincides with that of the diagonals of the load-bearing truss.

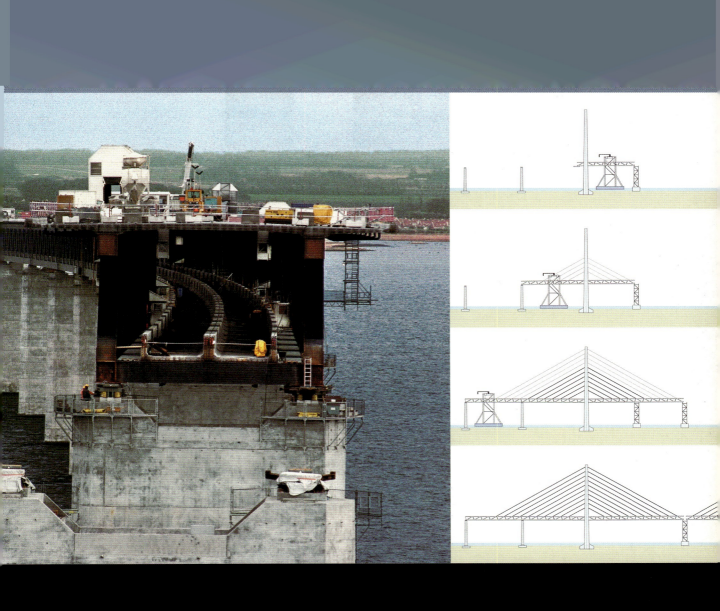

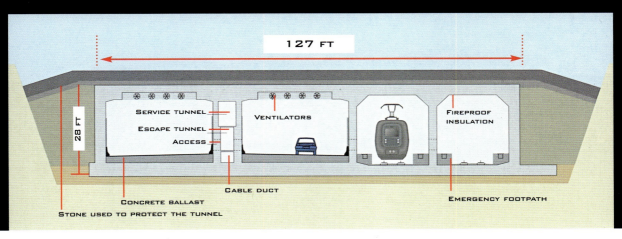

134-135 ■ The bridge has two railroad lines on the lower level and a wider road on the upper level.

135 top right ■ The plans are of four phases of construction and the position of the temporary supports in the main span.

135 center ■ The section between Copenhagen and the man-made intermediary island of Peberholm is formed by six tunnels: two for rail traffic, two for road traffic and two, smaller, for emergency use.

135 bottom ■ The construction phases and equipment of a pair of pylons are shown in detail.

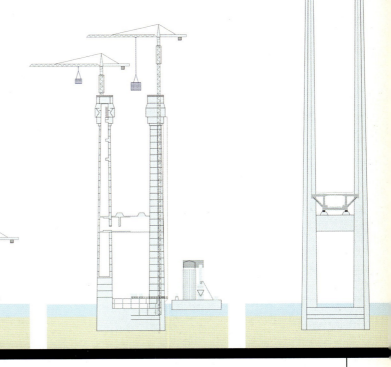

Location	Designed by	Length – Widest Span	Type	Construction
Pontevedra (Spain)	Leonardo Fernandez Troyano	423 ft 410 ft	Cable-stayed bridge with single pylon	1995

Puente Leréz

Leréz River, Pontevedra, Galicia (Spain)

The Leréz River Bridge was the result of the development of an original idea that had been implemented for the first time in the Sancho El Mayor Bridge in Navarra, Spain. In designing the Pontevedra Bridge over the Leréz, Leonardo Fernandez Troyano found himself faced by the same problem he had encountered more than ten years previously when designing the bridge over the Ebro River at Tudela: the asymmetry of the river bed had obliged him to find an absolutely new solution, and the result was a bridge with a single inclined pier, without spans on the river bank to create balance.

Leonardo Fernandez Troyano, who was born in Madrid in 1939, began working with his father Carlos but set up his own company in 1966 – *Oficina de Proyectos Carlos Fernandez Casado S.A.* All his bridges, built in Spain and Mexico, are original and expressive.

The Leréz Bridge was built in 1995 with a single cable-stayed span of 410 feet.

Three sets of stays radiate from the pier, which stands aligned with the

136 ■ Leréz Bridge has a single pylon whose section varies gradually from an approximate T-shape at the base to a trapezoidal form. Three bands of stays are attached to the pylon that together form an angle of 120°.

136-137 ■ The off-center direction of the pylon positioned on the bridge axis was determined by the static forces. The band of stays tied to the midway point of the road supports the deck, while the other two bands are anchored to the ground and act as counterweights.

road, creating an angle of 120°. One of the stay sets (the vertical one that runs down the road axis) supports the bridge's central span; the other two are anchored in counterweights in the flowerbeds in the traffic circle and traffic island of the roads that lead to the bridge.

The tip of the pylon stands 184 feet above the road surface. The three-dimensional nature of the structure is highlighted by the relative positions of the counterweight cables; those in the pylon ensure the stays form the lines of a hyperbolic parabola (in other words, they are not set on a flat surface). The aesthetic result of this geometrical design is very impressive.

The geometry of the structure (in particular the inclination of the pylon) is a consequence of the static balance created by the polygon of forces that act on the bridge.

Troyano studied the form and measurements of the pylon very carefully: the direction of the axis and the variation of the section along the axis (both in dimension and shape) are an elegant solution to the static and aesthetic requirements.

PUENTE LERÉZ

138 top, 139 top and center ■ The sketch, drawings and model illustrate the structural arrangement conceived by Troyano to overcome problems created by the asymmetry of the Lérez riverbed. Symmetry is created in the structure by a three-dimensional arrangement in which the support cables for the central span are balanced by another two bands, symmetrical with respect to the vertical plane through the bridge axis. The direction of the ground anchorages forces the cables to lie on a lined, not smooth, surface without disturbing the symmetry.

138 bottom ■ The night-lighting shows up the structure of the bridge to the observer about to cross. In the background the pylon rises like a lighthouse.

139 bottom ■ One end of the bridge stands on a large traffic island, with the pylon at the center. The anchorage blocks for the lateral stays stand on two median strips.

ERASMUS BRIDGE

NEW MAAS RIVER, ROTTERDAM (THE NETHERLANDS)

140 top ■ The north bank as seen from the cable-stayed bridge, showing the buildings that have remade the appearance of Rotterdam, which was devastated during World War II.

140 bottom and 140-141 ■ The complex that crosses the New Maas river in Rotterdam is formed by two bridges reached via a viaduct on the north bank. Officially the structure is named after the humanist Desiderius Erasmus, but it is known as the 'Swan' due to the elegance of the central section, pylon and smaller stayed span.

Construction of the bridge, which is named after Rotterdam's most famous son, was part of the Kop van Zuid development plan for a vast area on the south bank of the New Maas.

The competition was announced in 1989, and the winner was Ben van Berkel. The young Dutch architect's design won over the others for its structural characteristics and highly unusual aesthetics. The giant support pylon that stands alone in the middle of the river has stimulated the imagination of the city's inhabitants so much that it is known as the 'Swan'. The entire bridge is 877 yards long and includes an access viaduct on the north bank of the river. The bridge over the water is made from two different types of steel structure that are joined without any interruption.

From the north bank, where the old city lies, one sees the more visible part of the dual-span, asymmetrical Swan. The larger of the spans (932 feet) is cable-stayed, and the other (243 feet) is supported by piers. A very characteristic feature of the bridge is the unusual design of the sky-blue, 456-foot pylon. The lower section is like an upturned V inclined to the south (i.e., toward the smaller of the two spans). The vertical upper section is about 200 feet tall and provides an anchor for the 32 stays of the larger span. The counterweight is provided by two stays that anchor the pylon to the south foundation of the smaller span. Made from sheet steel, the

142 ■ The two stays that connect the pylon to the foundation common to both bridges are anchored high on the south face of the pylon.

143 top ■ The cycle and pedestrian lanes continue uninterrupted at the sides of the bridges, outside the pylons and stays.

143 bottom ■ The edges of the mobile part of the bridge run parallel to the direction followed by the ships and, in consequence, they cut through the decks obliquely.

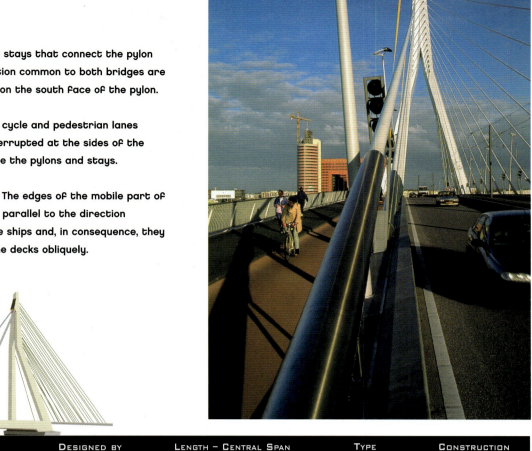

Location	Designed by	Length – Central Span	Type	Construction
Rotterdam (the Netherlands)	Ben van Berkel	2,631 ft / 932 ft	Cable-stayed bridge	1994-1996

pylon has a closed section and is strengthened by horizontal elements. An internal elevator allows maintenance inspections and operations to be carried out.

The second part of the Swan – a bascule bridge – is 400 feet long. It was included to permit river traffic to pass on a route that runs parallel to the south bank of the river. The section that can be raised is in the unusual form of an oblique parallelogram and revolves around a horizontal axis that is not perpendicular to the axis of the bridge. In addition to its unusual geometrical design, the size of the moving section is also exceptional (180 feet long by 125 feet wide). The movement is powered by an electro-hydraulic system that requires only a minute to open the bridge and a minute and a half to close it.

The deck is also made of steel. It is composed of 28 prefabricated elements, each 49 feet long and 125 feet wide. The road level has two traffic lanes divided by by a light-rail track; at the sides of the road there are two cycle and pedestrian lanes reached from the car park on the north side. The bridge is 41 feet above the water, so that users can see the passing flatboats and other vessels.

The Erasmus Bridge is the Port of Rotterdam's new symbol. It is made very attractive at night when lit up in profile by a lighting system also designed by Ben van Berkel.

144-145 ■ The greatest problem faced by the designer was the shape of the pylon, which appears in the picture like a giant tuning fork.

144 bottom ■ The deck is joined to the pylons by stays only: there is no other kind of connection in order to avoid the transmission of vibrations during earthquakes.

145 right ■ The height of the pylons (492 feet above the water) means that they can almost be seen from Lisbon city center.

Puente Vasco De Gama

Spanning the Tagus Estuary, Lisbon (Portugal)

The new crossing over the River Tagus in Lisbon was completed in 1998. Dedicated to the famous Portuguese explorer, the bridge is one of the several large-scale engineering works executed in the Iberian peninsula as part of the fair to mark the 500th anniversary of the discovery of America. The bridge is in three parts: a 7.6-mile section that crosses the water, and two access viaducts from the south and north, respectively 3.1 miles and 612 yards long. The bridge itself, which soars over the navigable section of the river, is a cable-stayed bridge 908 yards long in three spans, the central one of which is 1,378 feet in length.

The designer, Michel Virlogeux, was worried foremost by the threat of earthquakes. As these are frequent in that zone, he designed a very light deck that is completely independent of the pylons and that will suffer the minimum possible risk from seismic vibrations. This system had already been tested by Armando Rito in the Arade Bridge, also in Portugal.

To ensure the independence of the deck and pylons, Virlogeux studied their shapes in great detail. The reinforced-concrete H-pylons reach 492 feet above

PUENTE VASCO DE GAMA

146 top ■ The deck is connected on either side to the upper section of one of the two extensions of the pylon.

146 bottom ■ The photograph shows the bridge deserted just a few minutes before the opening ceremony on March 31, 1998.

147 ■ Enlargement of a detail in the structure reveals geometric patterns and effects of the light.

the water level; the two legs of each pylon are connected by two crosspieces: the first lies just above the water level, where the legs are at their widest apart, and the second (258 feet higher) lies at the level that the legs run upwards in parallel.

The deck runs over the water at a height of 164 feet, then passes between the legs of the pylons, though it is connected to them only by the cable-stays. The deck is of mixed structural type: two longitudinal, reinforced-concrete girders support the transversal steel supports, while in the central section the deck is supported completely by the cables. In the two lateral spans the girders each rest on three piers, which are formed by two parallel pillars.

The extremely long Vasco da Gama Bridge continues south over a marshy area where the river is very shallow. The deck is supported by piers identical to those in the lateral spans of the cable-stayed section.

Location	Designed by	Length – Widest Span	Type	Construction
Lisbon (Portugal)	Michel Virlogeux	2,723 ft 1,378 ft	Cable-stayed bridge	1995-1998

Suspension Bridges

▲ MILLENIUM   ▲ SOLFERINO   ▲ GATESHEAD MILLENNIUM

  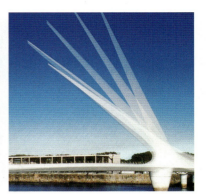

▲ Millennium   ▲ Solferino   ▲ De la Mujer

AT THE END of the 1980s a completely new interest arose in footbridges. Until then, they had been considered as no more than modest substitutes for road bridges, or as a function for an old, downgraded bridge that had become unsuitable for motorized traffic.

Today, the discovery is that city road bridges are inadequate for foot passengers: it is unlikely that a pedestrian will be able to take pleasure in stopping on a road bridge to admire the river. His enjoyment will be spoilt by the traffic and perhaps even by a little pushing and shoving by other pedestrians wishing to cross in a hurry.

The need to make modern metropolises enjoyable has prompted many European cities to build footbridges, and the most renowned engineering and architectural firms have entered this category of design competitions.

It might be said that a city footbridge is a new sort of bridge that gives great scope to the imagination of the designer but which creates restrictions through the inevitable aesthetic and functional requisites. A new footbridge has to suit its surrounding environment and must satisfy the many requirements of the pedestrians who will use it. In the last decade many excellent footbridges have been constructed, and though all different, they share certain characteristics: a lightness of design, an all-round view and dramatic lighting at night. Their designers are free to indulge their whims in the choice of materials and inventiveness of the structure, with the result that footbridges represent a sort of engineering extravagance.

Technology and aesthetics

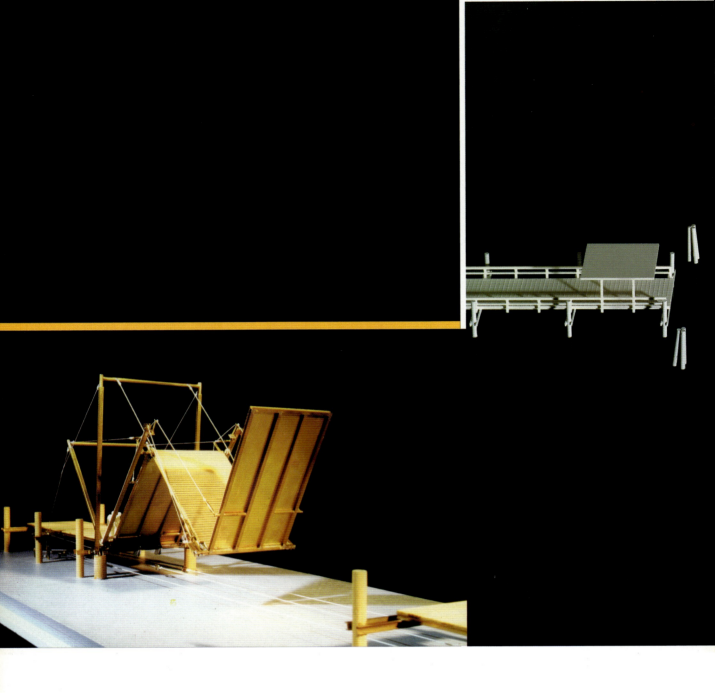

152 top and 152-153 bottom ■ Schlaich's study used models to perfect the mechanisms and control of the movements that had to be carried out while maintaining a statically determined configuration in any position; the outcome of the theoretical analysis and study of the models was to exclude certain elements, such as springs and hydraulic jacks, that may have caused unwanted movements.

152-153 top ■ The movement controlled by a single motor was transmitted to three winches that raised the three sections of the bridge through three pairs of cables.

Kiel Hörn Footbridge

Kiel (Germany)

The footbridge crosses the port of Kiel, the capital city of the Schleswig-Holstein region in north-central Germany. The bridge was part of the Hörn project, a large development plan for this seaport city, known to sailors the world over for its Kieler Woche, or Kiel Week (an annual series of international races held in the Aussenförde, the city's bay).

The footbridge was designed by the well-known firm of Schlaich Bergermann and Partner, famous for its ability to solve old problems with new concepts. Schlaich Bergermann has built many footbridges in Germany in the last decade, each of which has different and original characteristics, though all share an aesthetic

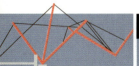

Location	Designed by	Length – Widest Span	Type	Construction
Kiel (Germany)	Jörg Schlaich, Vilkwin Marg	381 ft / 84 ft	Bascule bridge	1997

awareness and respect for their environment.

For the Kiel project, the firm produced a particular type of bascule bridge that has no precedent. The inspiration may have been military tanks that carry bridge structures which fold up on themselves; in this case, however, the structure opens and is transformed into a small bridge, but here the problem is on a different scale as the bridge had to be 380 feet long and 20 feet wide.

When the bridge is closed (i.e., open to foot traffic but closed to river traffic), it appears to be a common cable-stayed bridge, and only when it opens to river traffic does the originality of the design become apparent. The deck is divided into three parts by hinges and opens by bunching up like a pair of bellows. The opening operation is performed through mechanisms that wind up the traction cables until the bridge folds up on itself vertically.

154-155 ■ The various movements to lower the bridge after a boat had passed demonstrate the complexity of the mechanism, which had also to fold the balustrades of the parapet. Opened and closed about ten times a day, the bridge provides a passage across the port to within the city.

Naturally, such a design presented difficulties, but these were skillfully overcome. The structural calculations, which had to take into account the static, dynamic, and kinematic problems, were set after tests had been carried out on the behavior of the materials (often new) and the climatic conditions. The strength of the cables was calculated for every position of the bridge, under the most unfavorable load conditions, and taking into account wind factors. After these calculations had been completed, the bridge was subjected to stringent experimental trials on site.

The innovative opening and closing movement of Kiel footbridge has a certain symbolism that helps it blend naturally into its environment: the opening system resembles the raising mechanisms of the dockyard cranes and has a clear affinity with the life of the port, while the bridge's bright colors are a pleasant contrast to the leaden gray of the sea.

Kiel Hörn Footbridge

156 top ■ The various elements that form the structure and mechanisms are highlighted in color: the deck girders are in red, the parapet in yellow, and the pulleys and hinge mechanisms in yellow.

156 bottom ■ The folded structure takes up very little space, which reduces the risk of receiving overly strong buffeting from the wind.

156-157 and 157 bottom ■ The structure (in yellow) connects the two end elements of the deck and is formed by three parts joined by hinges. At top it is folded, and at bottom it is in the final, open position.

Solferino footbridge

River Seine, Paris (France)

The Solferino footbridge connects the Musée d'Orsay on the Quai Anatole France to the Tuileries Gardens on the Quai des Tuileries. Built between 1997 and 1999, it was created to fill the gap left by the demolition of a temporary footbridge that replaced the ancient Solferino Bridge from 1961 to 1992.

The lovely three-span, cast-iron bridge had lasted one hundred years. The name celebrated a military victory in 1859 at Solferino in Lombardy, Italy, when the French Emperor Napoleon III (an ally of Piedmont in the second war for Italian independence from the Austro-Hungarian Empire) defeated the Austrian Emperor Franz-Josef.

The modern footbridge crosses the Seine in a single span of 348 feet and forms an integral part of the urban landscape — a landscape in which the bridges perform a major role. The design by Marc Mimram solves the technical, aesthetic and city planning problems with complete originality; access to the bridge is given from four different points via a set of asymmetrical walkways that are cleverly inserted into the symmetry of the structure.

Structurally the bridge is very simple: there are two reticular caisson arches connected by crosspieces that support the

158 ■ Pedestrians descending the steps of the lower arch of the footbridge enter the underpass to the Tuileries. In the background we see the Pont Royal and, in the distance, Notre Dame.

158-159 ■ The two flights of steps that rise onto the arch from the two banks come out onto the wide upper platform in the middle of the bridge.

159 bottom ■ The slender, linear profile of the bridge is accentuated at night by the effect of the lights reflected in the water.

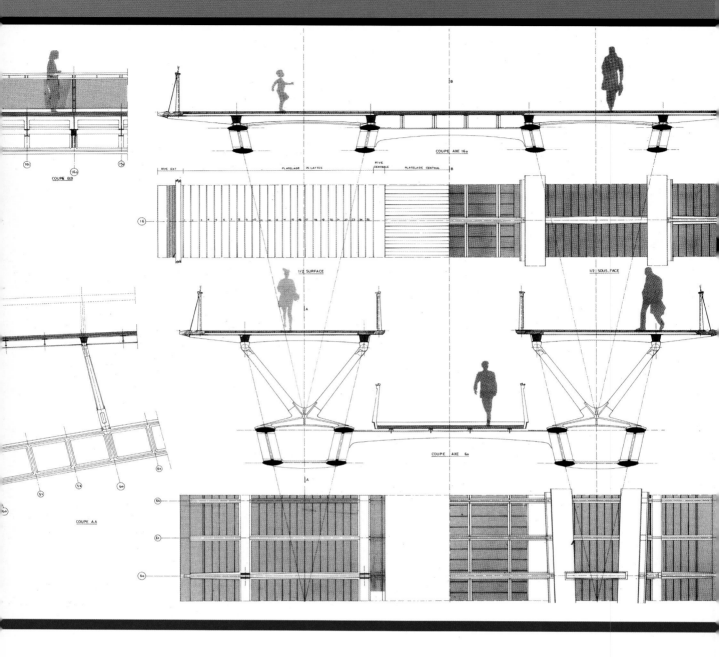

160-161 and 161 bottom ■ The plans and sections of the various levels included in the 'summary' of the project were carefully drawn to show how the different walkways are organized; they illustrate the alternatives offered by this innovative footbridge, on which it is possible to admire the view from anatomically shaped benches.

161 top and center ■ The upper platform that connects the Quai Anatole France to the Quai des Tuileries starts at the level of the two avenues and rises slightly toward the center. Here it meets the lower walkway formed by two flights of steps that descend on one side to the embankment parallel to Quai Anatole France, and on the other to the underpass to the Tuileries.

deck. The steel arches are blocked at the ends and have sections that decrease in size from the springers as they progress toward the crown. The deck is made from a mixture of steel and wood.

The inclusion of several walkways required a complex design. The problem lay in the need to offer access to the footbridge — which crosses the river in a direct line from the Quai Anatole France to the Quai des Tuileries — from the road that runs low down alongside the river, the left bank of the Seine, and from the underpass on the right bank that joins the Tuileries to the sidewalk at river level. Mimram had the idea of two stairways that climb the arch and, starting respectively from the left and right banks, reach the level of the deck, and come out in the middle of the bridge. Here the two walkways join the two straight sections that extend from the banks, including an intermediate section where benches and lampposts are installed. The walking surface of the upper walkway and the stairways are made from exotic wood with fine inlaid decorations.

The inclusion of the access walkways inside the structure does not alter the overall form of the footbridge, which appears complete, unencumbered and light.

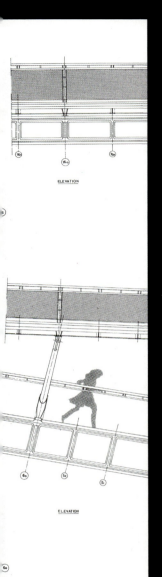

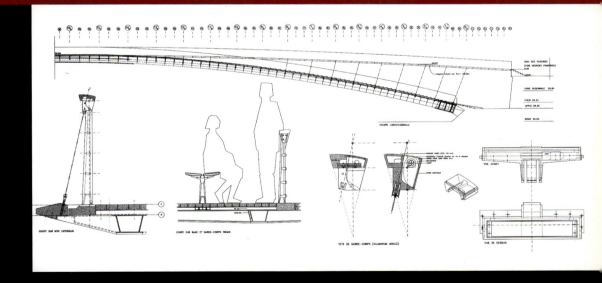

Location	Designed by	Length – Central Span	Type	Construction
Paris (France)	Marc Mimram	459 ft / 348 ft	Arch bridge	1997-1999

SOLFERINO

162-163 and 163 right ■ The suspension bridge is supported on either side by four cables attached to the tops of the V-shaped pylons. The bridge's unusual appearance is given by its flat and wide profile, which in turn creates an impression of lightness and extreme flexibility.

162 bottom ■ On the inauguration day, oscillations greater than forecast were measured, so that dampeners had to be added.

Millennium Footbridge

River Tyne, Newcastle (Great Britain)

The Millennium Bridge was built between 1996 and 2000 and is the only footbridge over the Thames. The new structure lies in the city's historical center, between Blackfriars and Southwark bridges. It connects the City (London's financial center) with Southwark, in a direct line between St. Paul's Cathedral and the new section of the Tate Modern Gallery, which is housed in a disused power station. The bridge was part of a renovation project for the old industrial district of Southwark.

The competition for the bridge was won by a group of famous names – the sculptor Sir Anthony Caro, architect Lord Norman Foster, and the engineering company Ove Arup & Partners – with a design that is original even to the eyes of the architecturally uneducated. Two features, important structurally, differentiate the London footbridge from common suspension bridges: the modest height of the support piers, which results in a tiny curvature of the support cables; and the position of the cables on the exterior of the bridge.

Overall the footbridge is 350 yards long. There are three spans suspended from two sets of four cables, each 5" thick. These are supported by two piers and are anchored on shore in blocks. The width of the central span is 472 feet and the suspension cables have a reduced rise of 7'6". The piers, which are very low, are formed by a reinforced-concrete base in the shape of a truncated cone, on which a wide, metal, V-shaped structure is positioned. The suspension cables run along the ends of the arms of the V (52 feet apart), and the cables hold up the deck supports. The supports are structural arms with a trapezoidal axis set at a distance of 26 feet from one another. These steel transversal arms support longitudinal supports, also metal, that form the walking surface of the deck. The parapets too are made from steel.

Apart from the low piers and very reduced curvature of the cables, the most apparent peculiarities of the bridge are the absence of stays (which usually join the deck to the suspension cables) and the large distance (6.5 feet) between the cables and parapet, itself only 13 feet wide.

The Millennium Bridge was opened to the public on June 10, 2000, but the unexpectedly large number of visitors, who crossed almost in rhythm, created oscillations greater than calculated and therefore the bridge had to be closed. In 2002, after a two-year period to review the calculations and insert dampers, the footbridge was reopened.

The modifications do not compromise the attractive looks of the bridge, which has a very slender and slightly arched profile. It is a respectful element of the river environment and designed, according to Norman Foster, to resemble 'a blade of light crossing the river.'

164 ■ The view from the Tate Modern Gallery highlights the expressiveness of the structure, which seems to lie quite naturally above the surface of the water.

164-165 ■ Crossing the bridge in line with St. Paul's cathedral, one has an all-round view of the city heightened by the fact that one is above the traffic.

165 bottom ■ The support cables are slightly curved and are connected directly to the transversal parts of the deck without the requirement for stays.

Location	Designed by	Length – Widest Span	Type	Construction
London (Great Britain)	Anthony Caro, Norman Robert Foster and Ove Arup & Partners	1,050 ft 472 ft	Suspension bridge	1996-2000

Gateshead Millennium Footbridge

River Tyne, Newcastle (Great Britain)

The Gateshead Millennium bicycle-and-pedestrian bridge was built in 2001 to connect Newcastle and Gateshead, two cities separated by the River Tyne but united by a common aspiration to become a fully qualified European cultural center. Crossing from Newcastle, one reaches the south bank of the river where the Baltic Center (a magnificent contemporary art exhibition area) and the Sage Gateshead (an international center for experimental music) are located.

The two cities have been nominated the European capitals of culture for the year 2008, and the new bridge will be the symbol of this status, not simply because it is a tangible expression of the uniqueness of the two cities, but because the bridge itself is a work of contemporary art, and the first that visi-

166 and 166-167 ■ The footbridge fits into the landscape discreetly but irresistibly. It is a slender, almost transparent structure that does not interrupt the view of the nearby bridges, but disguises them, thereby drawing attention to itself, its movement and extraordinary illumination.

tors will see when attending an exhibition at the Baltic Center.

As the Tyne is navigable, the bridge was not allowed to impede the river traffic. The problem was tackled by the Wilkinson Eyre architectural firm and Gifford & Partners engineering company, and resolved with the design for a bascule bridge of new conception. The structure is formed by two steel arches, one of which bears the load and supports the other with suspension cables. The two arches share springer points and the angle formed by the planes on which each arch lies is fixed. The whole structure revolves around the axis that unites the two hinges on either side of the bridge.

There are three characteristics that give the bridge its innovative appearance: the curved axis of the crossing surface, the deck hung from an

Location	Designed by	Length – Widest Span	Type	Construction
Newcastle (Great Britain)	Wilkinson Eyre, Gifford & Partners	413 ft 344 ft	Arch bascule bridge	2001

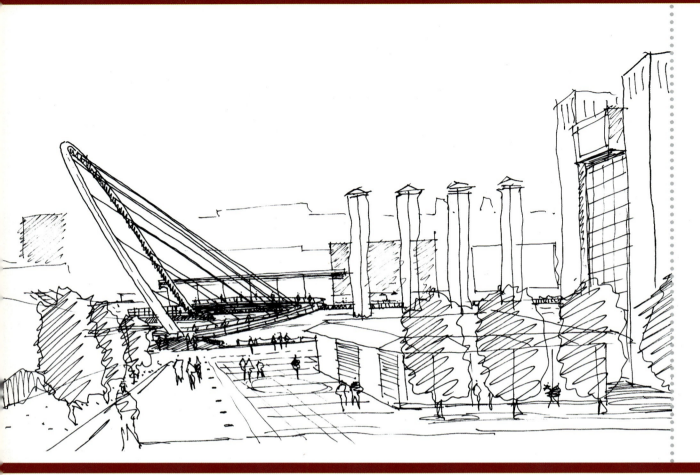

GATESHEAD MILLENNIUM FOOTBRIDGE

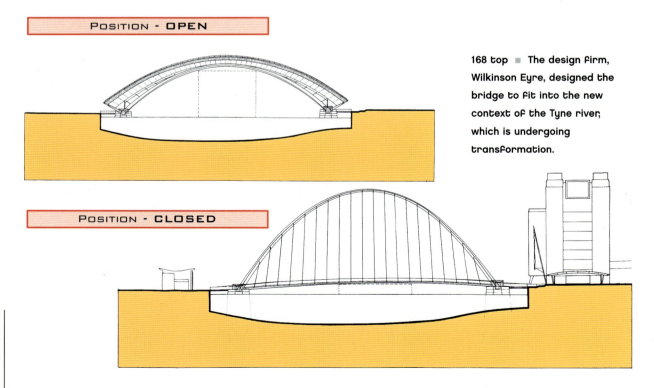

Position - Open

Position - Closed

168 top ■ The design firm, Wilkinson Eyre, designed the bridge to fit into the new context of the Tyne river, which is undergoing transformation.

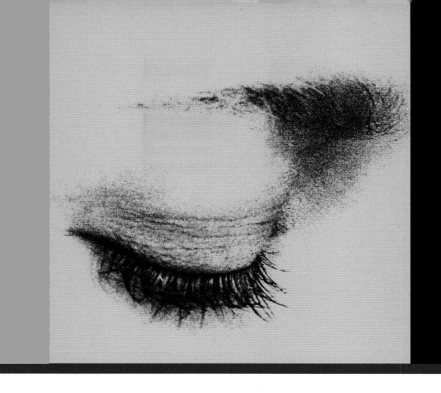

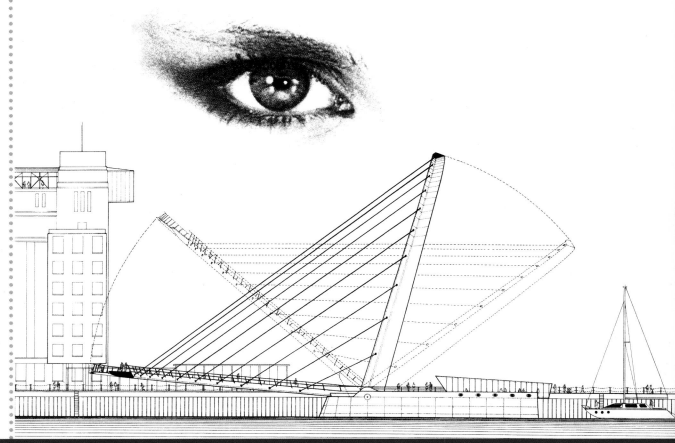

168 bottom ■ The opening mechanism of the bridge operates with the rotation of the entire structure around two pins at the end of the arch. The plan shows the two end positions.

169 ■ The opening and closing movement of the bridge has been dubbed the 'blinking eye.' The inhabitants of Newcastle have enthusiastically welcomed the innovative design of the cycle and pedestrian bridge, which allows them to cross directly to the new arts and entertainment area in Gateshead on the other side of the river.

oblique arch, and the unusual opening system (during movement the load-bearing element also revolves and the entire structure rotates around a longitudinal axis).

The overall length of the footbridge is 138 yards and the length of the span is 344 feet. The two lanes – for pedestrians and bicycles – are separate: the first lies on the inside of the curve and, for reasons of safety, is on a higher level.

The bridge opens in four minutes in response to a mechanism powered by eight electric motors. Each time it is opened (about 200 times a year), an automatic washing system is activated to clean the crossing surface and collect the rubbish in containers at either end of the arch.

The bridge was built on the premises of AMEC in Wallsend, six miles away by river, and was transported in two days to its operational location. To accomplish this, the world's largest floating crane, the Asian Hercules II, was used in combination with the high tide.

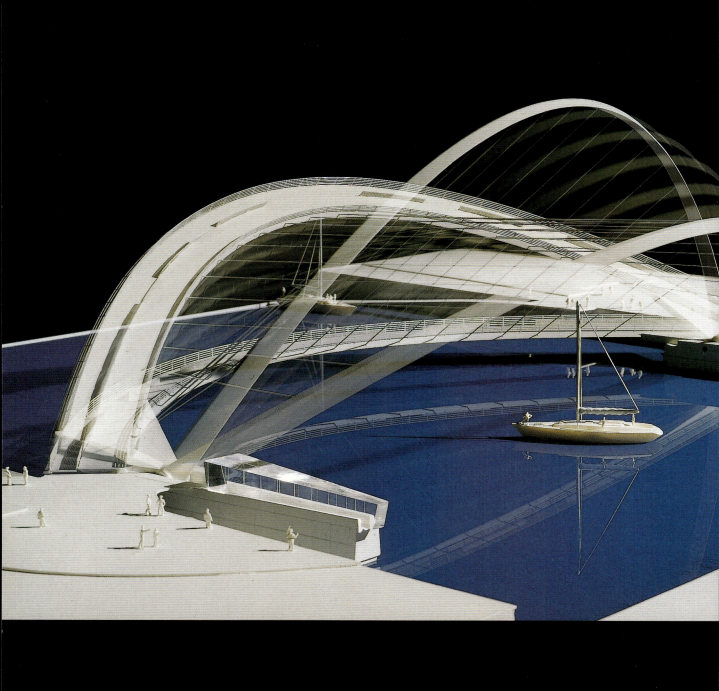
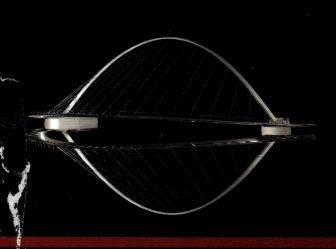
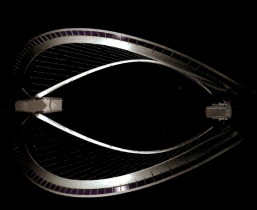

Although the bridge has been engineered to withstand the impact caused by a ship of 4,000 tons traveling at 4 knots, for further safety a barrier has been created in the water. This barrier prevents ships from approaching the springers and forces them to advance through a narrow strip in the center of the river.

The bridge's attractive appearance is enhanced by a computer-controlled lighting system that varies the colors of the lights and creates interesting reflections on the water.

Another unusual effect is produced by the innovative opening movement of the bridge, as a result of which it has been nicknamed the 'blinking eye.'

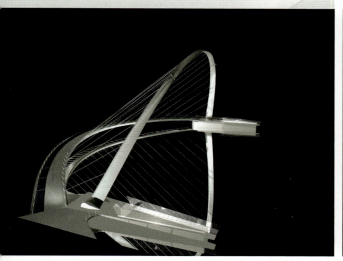

170-171 ■ The model was built to examine all the positions during opening and the full height achieved

170 bottom and 171 bottom ■ The computer-based study enabled analysis of the different configurations from different viewpoints. The reflection in the water was another visual aspect that was examined. To the observer watching from the side, the effect is like an eye half-closed or open, while a foreshortened view shows the geometric pattern of curves and hatched areas.

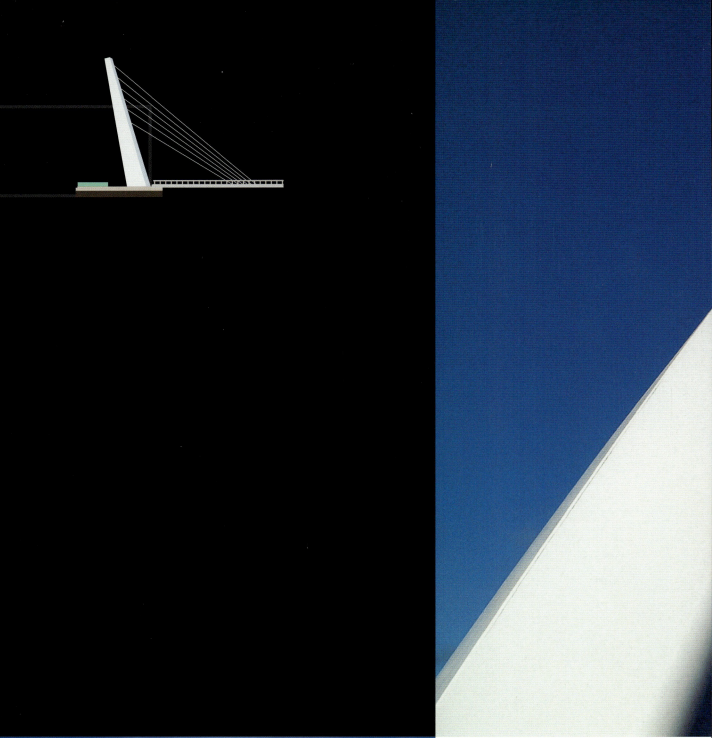

172-173 ■ The support arch is very slender and its three-dimensional effect seems the result of curving a flat, rectilinear band.

172 center and bottom and, 173 bottom ■ To a moving observer the open bridge appears in different forms, as is demonstrated by the three photographs. We see very precise figures: a V, a series of figure-eight shaped curves, and a heart.

GATESHEAD MILLENNIUM FOOTBRIDGE

174-175 ■ A computer-based lighting system draws attention to the bridge at night using different colors. Impressive effects are obtained using the reflections on the water.

La Mujer Footbridge

Puerto Madero, Buenos Aires (Argentina)

The district of Puerto Madero was recently redeveloped almost from scratch by the local Corporatión Antiguo Puerto Madero S.A. The plan created a new urban district with services, a road network, and parks. The proposal for a footbridge over the *dique* 3 of Madero port was put forward by entrepreneur Alberto Gonzales to connect Avenida Alicia Morenaude Justo with the east commercial area.

The Passerella De La Mujer was built in 2002. The bridge is 175 yards long with two fixed spans of 111 and 85 feet, and an intermediate cable-stayed moving span 328 feet long. The design was by Santiago Calatrava.

For ships to pass, the central section of the mobile span (approximately 218 feet wide) is able to rotate 90° around a white concrete pier 31 feet high that contains the rotation mechanism. The side piers, which are also made of white concrete, are in the form of round cylinders topped by elliptical capitals.

The deck is formed by a steel caisson girder strengthened by transversal and longitudinal plates.

The most important element both

176-177 and 177 bottom ■ The tall, elegant inclined pylon acts as a support to the main span via the cable stays. This design has kept the horizontal truss very light and slender and has reduced the disparity in weight between the two spans, thereby limiting as much as possible the flexion strain on the vertical axis of the rotation pin.

178 top and 178-179 ■ The three-dimensional models in the four pictures allowed the designers to observe the bridge from different standpoints and positions, to study and check its mechanisms and to observe its reflection.

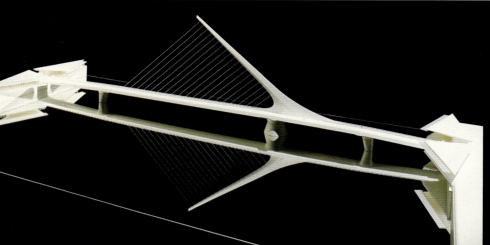

Location	Designed by	Length – Central Span	Type	Construction
Buenos Aires (Argentina)	Santiago Calatrava	525 ft 230 ft	Cable-stayed bridge	2002

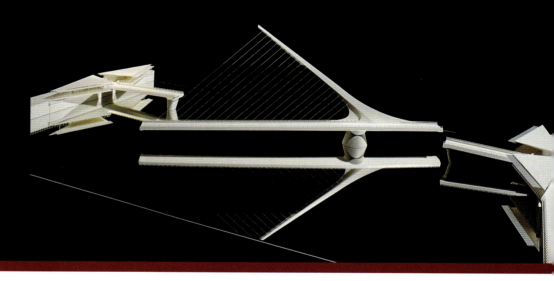

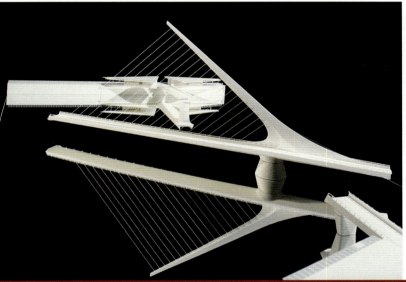 

178 bottom and 179 top ■ The positions of the bridge during the movement are shown in the two photographs, one of which shows Porto Madero (the old port in Buenos Aires as it is today, completely transformed by large new buildings, the Hilton Hotel, banks and the Sea Museum) and the other of the gardens that separate the buildings from the water's edge.

structurally and aesthetically is the steel pylon that rises 115 feet above the central pier (around which the mobile section rotates) and inclined at an angle of 38.81° to the horizontal. During opening of the bridge, the pylon rotates with the deck so that it faces west when the bridge is closed and north when it is open.

The components of the pylon are of different thickness but strengthened internally. Its elegance is created by the form of the section and its carefully calculated law of variation.

The various parts of the metal structure were produced in Spain and assembled on site.

INDEX

Note   c = caption

A

Aelius Bridge, 16
Akashi Strait, 116, 118
Akashi Kaikyo, bridge, 116, 116c, 117, 118, 119, 119c, 129
Alboregas, river, 102
Alby, Amédée, 62, 64c
Alcántara Bridge, 18, 18c, 19
Alexander III, czar, 63
Alexandre III, Pont, 26c, 62, 63, 63c
Allegheny Aqueduct , 40
Allonzier-La Caille, 30
Amman, 97
Anglesey, island, 28, 28c, 29, 34
Angostura Bridge, 92
Anji Bridge, 18c, 19
Annecy, 31c
Annecy, Pont d', 30
Antonio da Ponte, 20
Aosta, 18
Apure, river, 92
Arade, bridge, 145
Arenas de Pablo, Juan J., 105, 107c
Arno, river, 20
Avignon, Pont d', 19
Avignon, 19, 19c
Awaji, island, 116, 118

B

Bajo Grande, port, 93c
Baker, Benjamin, 57, 58, 58c
Beijing, 19
Belin, E., 31
Berlin, 40
　Royal Polytechnic Institute, 40
Bern, 72
Bernini, Gian Lorenzo, 18
Berthier, Ch., 31
Bertin, E., 31
Betacourt, Romulo, 92
Blackfriars Bridge, 163
Bonnardet et Blanc, company, 31
Bradfield, John Job Crew, 74, 76, 77c
Bristol, 32
Britannia Bridge, 34, 34c
British South Africa Company, 70
Brooklyn, 39, 41c, 42c, 97
Brooklyn Bridge, 10c, 25c, 26c, 32, 39, 39c, 42c, 45, 84
Brown, Samuel, 29
Buchanan Eads, James, 36, 36c, 37
Budapest, 32, 32c, 33
Buenos Aires, 176, 178c, 179c
　Hilton Hotel, 179c
　Sea Museum, 179c
Bureau des Dessinateurs, 26

C

Caille, Pont de la, 30, 31c
Cairo, 80
Calatrava, Santiago, 102, 103c, 178c
Cantal, 48, 49c
Capetown, 70
Caquot, Pont, 31
Caro, Sir Anthony, 163, 164
Cassiet, Bernard, 62
Castel Sant'Angelo, 18
Charles V, 19
Charles-Albert, Pont, 30
Charles-Albert of Savoy, 30
Chicago, 36
Cincinnati, 40, 42
Clarke, R., 84c
Coalbrookdale, 26
Copenhagen, 133, 133c, 135c
Cornwall, 16, 53
Corps des Ponts et Chaussées, 26
Corporation Antiguo Puerto Madero S. A., 176
Constantine, emperor, 16
Coulant, Jules, 64
Cousin, Gaston, 62
Cruselles, 30

D

Danube, river, 32, 33c
De la Mujer, footbridge, 150c, 176
Dogali Bridge, 68
Dongguan, 124
Döring, Anselm, 90
Dorman Long & Co., company, 74
Douro, river, 48
Doyelle, François, 109c
Dublin, 28
Dungan, C., 84

E

Eads Bridge, 25c, 36, 37
East Bridge, 129
East River, 39, 39c, 42c
Ebro, river, 136
Ecole de Marine, 26
Ecole des Mines, 26
Ecole des Ponts et Chaussées, 26
Ecole Polytechnique, 10, 26
Edinburgh, 57, 58c
Eiffel Alexandre-Gustave, 48, 48c, 49c
Emerita Augusta, 102
Erasmus, *see* van Rotterdam
Erasmus Bridge, 140, 143
Espalion, 19
Estaing, 19
Estremadura, 102

F

Fairbarn, William, 34, 35, 35c
Feuerthalen, 97
Filtrannan, channel, 133
Firth of Forth, 57, 57c
Florence, 20, 20c
Forth Rail Bridge, 25c, 58
Foster, Lord Norman Robert, 163, 164, 164c
Fowler, John, 57, 58c
Franz-Josef Hapsburg, 158
Freeman, John R., 70c, 71, 74, 77c
Frémiet, Emmanuel, 64
Fremont, John C., 83
Freyssinet, Eugène, 90
Funen Island, 126, 126c, 127, 128c, 129c

G

Gaddi, Taddeo, 20
Garabit Viaduct, 48, 48c, 49
Gateshead, 166, 169c
Gateshead Millenium, footbridge, 149c, 166, 167
Gateway Arch, 37c
General Paez, Puente, 92
General Urdaneta, Puente, 92
George Washington Bridge, 99
Gerber, Enrico, 57
Giang-Tung-Giao, bridge, 16
Gifford & Partners, firm, 167, 168c
Geneva, 72
Giotto, 20
Girard, Brice, 109c
Golden Gate Strait, 83, 83c
Golden Gate Bridge, 10c, 67c, 68, 83, 83c, 84, 99
Gonzales, Alberto, 176
Grand Belt Bridge, 90c
Granet, Pierre, 64
Graubünden, 72, 72c
Great Belt, East Bridge, 118
Great Belt Link, 126, 126c, 127
Greenwich, 116
Guadalquivir, 105

INDEX

Guiana, 92
Guardiana, river, 102

H
Hadrian, Aelius, emperor, 16
Hadrian's Bridge, 16
Hakata Oshima Bridge, 117
Harbour Bridge, 67c, 74, 74c
Hartman Bridge, 10c
Harvey, Bill, 100c, 101
Haute Savoie, 30, 31c
Hennebique, François, 68, 72
Henri IV, 20
Herrmann Amman, Othmar, 97, 99c
Hitsuishjima-Seto Bridge, 116c, 117
Holyhead, port, 28
Honfleur, 108
Hong Kong, 122c, 123, 124c, 125
Honshu, 116, 117c, 118
Honshu-Shikoku Bridge Authority, 118c, 120c
Houston, 10c
Humber, river, 101
Humber Bridge, 101, 118
Hylan, John F., 97

I
Ikuchi Bridge, 117
Ikuchijima, island, 120
Imbault, George Camille, 71
Innoshima Bridge, 116, 117
Invalides, Pont des, 62
Istanbul, 83
Iwakurojima-Seto Bridge, 116c, 117

J
Jones, Sir Horace, 51, 53, 53c

K
Kap Shui Mun Bridge, 122c, 123
Keystone Bridge Company, 36
Kiel, port, 153, 154
Kiel Horn, footbridge, 153, 154, 155
Kingston on Hull, 100c, 101
Kita Bisan-Seto Bridge, 117, 117c
Könen, Gerd, 90
Kowloon, island, 123
Kurushima Strait, 117
Kurushima Kaikyo Bridge, 116c, 117

L
La Barqueta, Puente, 104c, 105
La Cartuya, island, 104c, 105
  Parco Isla Mágica, 104c, 105
  Parco Sevilla Tecnopolis, 104c
Landau, island, 123
Lavigne, Charles, 109c
Le Corbusier (Charles-Edouard Jeanneret), 13
Le Havre, 10c, 108, 109c, 112
Lehaître, F. P., 31, 31c
Lenoir, Alfred, 64
Leon, Pilin, 93c
Lérez, river, 136, 139c
Lérez, Puente, 136, 137
Li Chun, 19
Lisbon, 145, 147c
London Bridge, 51
London, 10c, 28, 32, 51, 51c, 53c, 163, 164c
  St. Paul's cathedral, 163, 164c
  Tate Modern Gallery, 163, 164c
  Tower of London, 51c
Lusitania, Puente, 102, 103, 103c
Lys, torrent, 18

M
Ma Wan, island, 123
Madrid, 136
Maillart, Robert, 72, 72c, 73, 73c
Malmo, 133, 133c

Manhattan, 39, 39c, 40, 42c, 97c
Manhattan Bridge, 42c
Maracaibo, 92, 92c, 93
Maracaibo Bridge, 92
Marg, Vilkwin, 154c
Marqueste, Laurant, 64
Maxentius, emperor, 16
Menai Bridge, 28c
Menai Straits, 28, 34
Mérida, 102, 103c
  Arch of Tiberius, 102
  Tomb of the Julii, 102
  Tomb of the Voconii, 102
  Temple of Diana, 102
  Temple of Mars, 102
Mérida, bridges, 18
Mézière 26
Michel, Gustave, 64
Middlesbrough, 74
Millenium, footbridge, 149c, 150c, 163, 164
Milvian Bridge, 16
Mimram, Marc, 158, 160, 161c
Minami Bisan-Seto, bridge, 117
Mississippi, river, 36
Missouri, river, 36
Monier, Joseph, 68, 72
Monkwearmount, 26
Morandi, Riccardo, 92, 92c, 93
Morrow, Irving F., 83, 84c
Mosi-oa-Tunya, park, 70
Mott MacDonald and Yee Associates, 124c

N
Napoleon III, 158
Naruto Strait, 116
Neuf, Pont, 13c, 20, 20
Neuvial Viaduct, 49
Neva, river, 64, 64c
New Maas, river, 140c, 141
New York, 10c, 39, 40, 76, 97, 97c, 99
  Statue of Liberty, 48c, 76
New York University, 97
New York Bridge Company, 39
Newcastle, 166, 168c, 169c
  Baltic, 166, 167
  Sage Gateshead, 166
Nicholas II, czar, 63
Normandy Bridge, 10c, 108, 110, 120

O
Octavian Augustus, emperor, 16
Oficina de Proyectos Carlos Fernandez
  Casado S. A., 136
Ohnaruto Bridge, 116, 116c
Omishima, island, 120
Omishima Bridge, 116c, 117
Øresund, 133
Øresund Strait, 134c
Øresund Bridge, 133
Orinoco, river, 92
Otto, emperor, 16
Ove Arup & Partners, 163, 164c

P
Pacific Ocean, 83
Paine, C. E., 84c
Palladio, Andrea, 16
Pantaleon, Marcos J., 105, 107c
Paris, 13c, 20, 20c, 26, 48, 62, 64, 158, 161c
  Champs-Elysées, 64
  des Invalides, 64, 64c
  Eiffel Tower, 48, 48c
  Louvre, 20
  Musée d'Orsay, 158
  Notre Dame, 159c
  Notre Dame, island, 20
  Saint German des Près, abbey, 20
  Tuileries, gardens, 158, 159c, 160, 160c

INDEX

Parramatta, river, 74, 74c
Peberholm, island, 135c
Penmon, 35
Pennsylvania Military Academy, 97
Perronet, Jean-Rodolphe, 26
Pittsburgh, 40
Pont Alexandre III, see Alexandre III, Pont
Pont Charles-Albert, see Charles-Albert, Pont
Pont des Invalides, see Invalides, Pont des
Pont Neuf, see Neuf, Pont
Pont Royal, see Royal, Pont
Pont Saint Bénezet, see Saint Bénezet, Pont
Pont Saint-Martin, see Saint-Martin, Pont
Pont vieux sur le Lot, 19
Ponte Sant'Angelo, see Sant'Angelo, Ponte
Ponte Vecchio, 20, 20c
Pontevedra, 136, 136c
Portland, 53
Porto, 48
Post Bridge, 16
Puente General Paez, see General Paez, Puente 92
Puente General Urdaneta, see General Urdaneta, Puente
Puente La Barqueta, see La Barqueta, Puente
Puente Lerez, see Lerez, Puente
Puente Lusitania, see Lusitania, Puente
Puente Vasco de Gama, see Vasco de Gama, Puente
Puerto Madero, 176, 179c
Pümpin et Herzog, firm, 72
Punta Davis, 74
Punta Wilson, 74

Q
Quebec, 58
Queens Bridge, 42c
Queensboro, 42c

R
Résal Jean, 62, 64c
Rhodes, Cecil John, 70
Rhone, river, 19
Rialto Bridge, 13c, 20, 20c
Rito, Armando, 145
Ritter, Wilhelm, 97
Roberts, Gilbert, 100c, 101
Röbling, John Augustus, 39, 41c, 42
Röbling Washington, 42, 45
Rome, 13c, 16
   Porta del Popolo, 16
Roosevelt, Franklin Delano, 83
Roth, Georg, 134c
Rotterdam, 140, 140c, 141, 143, 143c
Royal, Pont, 159cd

S
Saint Bénezet, Pont, 19, 19c
Saint-Flour, 48
Saint Louis, 36, 36c, 37c
Saint-Martin, Pont, 18, 18c
Salgina, Valle 72, 73
Salginatobel Brücke, 72, 73
Salvio Giuliano, 16
San Francisco, 83, 83c, 84c
San Francisco Bay, 83
San Francisco-Oakland Bridge, 83
San Lorenzo Bridge, 58
Sancho El Major, bridge, 136
Sant'Angelo, Ponte, 13c, 16
Schiers, 72
Schlaich, Jörg, 154
Scuders, 72
Seine, river, 20, 62, 63, 64, 108, 160, 160c
Septimius Severus, emperor, 16
Seto-Ohassi Bridge, 117
Seto Sea, 116, 117, 120, 121c

Severn, river, 26
Shikoku, 116, 117c
Shimotsui-Seto Bridge, 117, 117c
Shin Onomichi Bridge, 117, 117c
Seoul Viaduct, 49
Seville, 105, 107c
Solferino, 158
Solferino, footbridge, 149c, 150c, 158
Solferino Bridge, 158
Southwark, 163
Southwark Bridge, 163
Stephenson, Robert, 35, 35c
Sprogo, island, 126c
State Island, 97c
Steiner, Clément, 64
Strass, Joseph B., 83, 84c
Sunderland, 26
Sydney, 74, 74c, 77c
   Opera House, 74c
Széchenyi Bridge, 32, 33, 33c

T
Tago, river, 18, 145
Tatara Bridge, 110, 117, 117c, 120, 121
Tavanasa, 73
Tay, river, 57, 57c
Telford, Thomas, 28, 29, 29c
Tiber, river, 16, 18
Thames, river, 51, 51c, 163
Thierney Clark, William, 32, 32c
Thomas, John, 35
Tower Bridge, 10c, 26, 51, 51c, 54
Trajan, emperor, 18, 19
Troyano, Carlos, 136
Troyano, Leonardo Fernandez, 136, 136c, 139c
Truyère, river, 48
Tsing Ma Bridge, 122c, 123, 125
Tsing Yi, island, 123
Tutela, 136
Tyne, river, 166, 167, 168c

U
Usses, torrent, 30, 31c

V
Van Berkel, Ben, 141, 143c
Van Rotterdam, Erasmus, 140c
Vasco de Gama, Puente, 90c, 145, 146
Venice, 13c, 20, 20c
   Grand Canal, 20
Verrazzano, Giovanni da, 99
Verrazzano Narrows, 97, 99c
Verrazzano-Narrows Bridge, 90c, 97, 97c, 99
Victoria Falls, 70, 70c, 71, 71c
Victoria Falls Bridge, 67c, 70
Virlogueux, Michel, 109c, 145, 147c
Vitellius, emperor, 16

W
Wallsend, 171
Watanabe, Kaichi, 58
Wear, river, 26
Wilkinson Eyre, firm, 167, 168c
Williamsburg Bridge, 42c
Wolfe-Barry, Sir John, 51, 53, 53c, 54

X
Xiao, river, 19

Z
Zambesi, river, 70
Zealand, island (Denmark), 126c, 127, 128c, 129c
Zhaozhou, 18c
Zimbabwe, 70
Zulia, 92
Zuog, 73
Zurich, 72, 97, 102

PHOTOGRAPHIC CREDITS

Page 1 Jean Gaumy/Magnum Photos/Contrasto
Pages 2-3 Jon Hicks/Corbis/Contrasto
Pages 4-5 Bassignac Gilles/Gamma/Contrasto
Pag. 6-7 Hulton-Deutsch Collection/Corbis/Contrasto
Pag. 9 and page 67 center Alamy Images
Page 11 Antonio Attini/Archivio White Star
Page 14 left Livio Bourbon/Archivio White Star
Page 14 right Marcello Bertinetti/Archivio White Star
Pages 14-15 Marcello Bertinetti/Archivio White Star
Page 16 left Giulio Veggi/Archivio White Star
Page 16 right Marcello Bertinetti/Archivio White Star
Page 17 Tommaso Bonaventura/Contrasto
Page 18 left Panorama Stock
Page 18 center Antonio Attini/Archivio White Star
Page 18 right Andrea Jemolo/Corbis/Contrasto
Page 19 Antonio Attini/Archivio White Star
Page 20 left Marcello Bertinetti/Archivio White Star
Page 20 right Marcello Bertinetti/Archivio White Star
Page 21 Livio Bourbon/Archivio White Star
Pages 22-23 Alamy Images
Page 25 left and page 60-61 top Colin Garratt; Milepost 92 1/2/Corbis/Contrasto
Page 25 center and page 36-37 Kelly-Mooney Photography/Corbis/Contrasto
Page 25 right and page 44 Adam Woolfitt/Corbis/Contrasto
Page 26 left and page 54-55 Marcello Bertinetti/Archivio White Star
Page 26 center and page 32-33 Carmen Redondo/Corbis/Contrasto
Page 26 right and page 64 Raphael Gaillarde/Gamma/Contrasto
Pages 28-29 Jason Hawkes/Corbis/Contrasto
Page 29 top Hulton-Deutsch Collection/Corbis/Contrasto
Page 29 bottom Topham Picturepoint/ICP
Pages 30-31 Alain Gaymard
Page 31 Jacques Mossot
Page 32 Marcello Bertinetti/Archivio White Star
Page 33 Barry Lewis/Corbis/Contrasto
Page 34 left Archivio Scala
Page 34 right Warren Kovach
Page 35 top left Hulton Archive/Laura Ronchi
Page 35 top right Hulton Archive/Laura Ronchi
Page 35 bottom Warren Kovach
Page 36 Corbis/Contrasto
Page 37 Antonio Attini/Archivio White Star
Page 38 Antonio Attini/Archivio White Star
Page 39 Alamy Images
Page 40 top Corbis/Contrasto
Page 40 bottom Corbis/Contrasto
Pages 40-41 Hulton Archive/Laura Ronchi
Page 41 Corbis/Contrasto
Pages 42-43 Antonio Attini/Archivio White Star

Page 43 Ron Watts/Corbis/Contrasto
Pages 44-45 Kelly-Mooney Photography/Corbis/Contrasto
Pages 46-47 Setboun/Corbis/Contrasto
Pages 48-49 Arthur Thévenart/Corbis/Contrasto
Page 49 top Science Photo Library/Grazia Neri
Page 49 bottom Corbis/Contrasto
Page 50 Marcello Bertinetti/Archivio White Star
Page 51 Marcello Bertinetti/Archivio White Star
Page 52 Hulton Archive/Laura Ronchi
Pages 52-53 Hulton Archive/Laura Ronchi
Page 53 top Hulton Archive/Laura Ronchi
Page 53 bottom Archivio Scala
Page 54 Marcello Bertinetti/Archivio White Star
Pages 54-55 Marcello Bertinetti/Archivio White Star
Page 55 left Marcello Bertinetti/Archivio White Star
Page 55 right Angelo Hornak/Corbis/Contrasto
Page 56 Colin Garratt; Milepost 92 1·2/Corbis/Contrasto
Page 57 Jason Hawkes
Page 58 top Alamy Images
Page 58 bottom Hulton Archive/Laura Ronchi
Pages 58-59 Hulton Archive/Laura Ronchi
Page 59 bottom Hulton Archive/Laura Ronchi
Pages 60-61 bottom Grant Smith/Corbis/Contrasto
Pages 62-63 Setboun/Corbis/Contrasto
Pages 64-65 Robert Holmes/Corbis/Contrasto
Page 65 Hrtmut Krinitz/Hemisphere
Page 67 left and page 80-81 Giulio Veggi/Archivio White Star
Page 67 right and page 70-71 Brian A. Vikander/Corbis/Contrasto
Page 68 left and page 88 bottom Morton Beebe/Corbis/Contrasto
Page 68 right kindly provided by Angia Sassi Perino
Page 70 Alamy Images
Page 72 Earthquake Engineering Library, University of California, Berkeley
Pages 72-73 Earthquake Engineering Library, University of California, Berkeley
Pages 74-75 Australian Picture Library
Page 76 Australian Picture Library
Pages 76-77 Australian Picture Library
Page 77 top Hulton Archive/Laura Ronchi
Page 77 bottom Australian Picture Library
Page 78 top Glenn Beanland/LPI
Page 78 bottom Australian Picture Library
Pages 78-79 Australian Picture Library
Page 82 Marcello Bertinetti/Archivio White Star
Page 83 Antonio Attini/Archivio White Star
Page 84 Underwood & Underwood/Corbis/Contrasto
Pages 84-85 Bettmann/Corbis/Contrasto
Page 85 Bettmann/Corbis/Contrasto

Page 86 Marcello Bertinetti/Archivio White Star
Pages 86-87 Roger Ressmeyer/Corbis/Contrasto
Page 87 Marcello Bertinetti/Archivio White Star
Page 88 top Macduff Everton/Corbis/Contrasto
Pages 88-89 Amos Nachoum/Corbis/Contrasto
Page 90 left and page 98-99 Will & Deni McIntyre/Corbis/Contrasto
Page 90 center and page 147 Alamy Images
Page 90 right and page 130-131 Gamma/Contrasto
Page 92 AFP/De Bellis
Pages 92-93 Yann Arthus-Bertrand/Corbis/Contrasto
Pages 94-95 Yann Arthus-Bertrand/Corbis/Contrasto
Page 96 Falke/laif/Contrasto
Page 97 Karl Lang
Page 98 Peter J. Eckel/Time Life Pictures/Getty Images/Laura Ronchi
Page 99 Bettmann/Corbis/Contrasto
Page 100 Jason Hawkes
Page 101 Alamy Images
Page 102 kindly provided by Santiago Calatrava S.A
Pages 102-103 kindly provided by Santiago Calatrava S.A
Page 103 top kindly provided by Santiago Calatrava S.A
Page 103 kindly provided by Santiago Calatrava S.A
Page 104 Benoit Decout/REA/Contrasto
Pages 104-105 Peter M. Wilson/Corbis/Contrasto
Page 105 kindly provided by Arenas & Asociados
Page 106 top kindly provided by Arenas & Asociados
Page 106 bottom kindly provided by Arenas & Asociados
Pages 106-107 Benoit Decout/REA/Contrasto
Page 107 kindly provided by Arenas & Asociados
Pages 108-109 Jean Gaumy/Magnum Photos/Contrasto
Page 110 top Bertrand Rieger/Hemisphere
Page 110 bottom Grant Smith/Corbis/Contrasto
Page 111 Jean Gaumy/Magnum Photos/Contrasto
Pages 112-113 Bassignac Gilles/Gamma/Contrasto
Page 113 top Jean Gaumy/Magnum Photos/Contrasto
Page 113 bottom Jean Gaumy/Magnum Photos/Contrasto
Pages 114-115 Bassignac Gilles/Gamma/Contrasto
Page 115 Jean Gaumy/Magnum Photos/Contrasto
Page 116 top left kindly provided by Honshu Shikoku Bridge Autority
Page 116 top right Kurita Kaku/Gamma/Contrasto
Page 116 center left kindly provided by Honshu Shikoku Bridge Autority
Page 116 center right kindly provided by Honshu Shikoku Bridge Autority
Page 116 bottom left kindly provided by Honshu Shikoku Bridge Autority
Page 116 bottom right kindly provided by Honshu Shikoku Bridge Autority

PHOTOGRAPHIC CREDITS

Pages 116-117 kindly provided by Honshu Shikoku Bridge Autority
Page 117 top left kindly provided by Honshu Shikoku Bridge Autority
Page 117 top right kindly provided by Honshu Shikoku Bridge Autority
Page 117 center left kindly provided by Honshu Shikoku Bridge Autority
Page 117 center right kindly provided by Honshu Shikoku Bridge Autority
Page 117 bottom Angelo Colombo/Archivio White Star
Page 118 kindly provided by Honshu Shikoku Bridge Autority
Pages 118-119 Kyodo News
Page 120 right kindly provided by Honshu Shikoku Bridge Autority
Page 120 left Kyodo News
Page 121 kindly provided by Honshu Shikoku Bridge Autority
Page 122 Yang Liu/Corbis/Contrasto
Pages 112-123 Michael S. Yamashita/Corbis/Contrasto
Page 123 Michael S. Yamashita/Corbis/Contrasto
Page 124 Michael S. Yamashita/Corbis/Contrasto
Pages 124-125 Michael S. Yamashita/Corbis/Contrasto
Page 125 top Derek M. Allan; Travel Ink/Corbis/Contrasto
Page 125 bottom kindly provided by Gleitbau Ges.m.b.H
Pages 126-127 top Hornbak Teit/Corbis Sygma/Contrasto
Pages 126-127 bottom Amet Jean Pierre/Corbis Sygma/Contrasto
Page 128 top left Amet Jean Pierre/Corbis Sygma/Contrasto
Page 128 bottom left Xavier Rossi/Gamma/Contrasto
Page 128 right kindly provided by Dissing+Weitling
Page 129 left kindly provided by Dissing+Weitling
Page 129 right Xavier Rossi/Gamma/Contrasto
Pages 130-131 Xavier Rossi/Gamma/Contrasto
Page 131 Karl Lang
Page 132 kindly provided by Øresunsbro Konsortiet
Page 133 kindly provided by Øresunsbro Konsortiet
Page 134 top kindly provided by Øresunsbro Konsortiet
Page 134 bottom kindly provided by Øresunsbro Konsortiet
Pages 134-135 kindly provided by Øresunsbro Konsortiet
Page 135 right kindly provided by Øresunsbro Konsortiet
Page 135 center kindly provided by Øresunsbro Konsortiet

Page 135 bottom kindly provided by Øresunsbro Konsortiet
Page 136 kindly provided by Studio Carlos Fernandez Casado
Pages 136-137 kindly provided by Studio Carlos Fernandez Casado
Page 138 top left kindly provided by Studio Carlos Fernandez Casado
Page 138 top right kindly provided by Studio Carlos Fernandez Casado
Page 138 bottom kindly provided by Studio Carlos Fernandez Casado
Page 139 top kindly provided by Studio Carlos Fernandez Casado
Page 139 center kindly provided by Studio Carlos Fernandez Casado
Page 139 bottom kindly provided by Studio Carlos Fernandez Casado
Page 140 top Alamy Images
Page 140 bottom Karl Lang
Pages 140-141 Alamy Images
Page 142 Karl Lang
Page 143 top Annebicque Bernard/Corbis Sygma/Contrasto
Page 143 bottom Karl Lang
Page 144 Benoit Decout/REA/Contrasto
Pages 144-145 Alamy Images
Page 145 Benoit Decout/REA/Contrasto
Page 146 top Marta Nascimento/REA/Contrasto
Page 146 bottom Bassignac Gilles/Gamma/Contrasto
Page 147 Alamy Images
Page 149 left and page 165 Peter Cook/VIEW
Page 149 center and page 158-159 Cyril Delettre/REA/Contrasto
Page 149 right and page 172-173 Alamy Images
Page 150 left and page 162 Alamy Images
Page 150 center and page 161 center Marc Mimram Ingenierie S.A.
Page 152 top kindly provided by Von Gerkan, Marg und Partner
Page 152 bottom kindly provided by Von Gerkan, Marg und Partner
Pages 152-153 kindly provided by Von Gerkan, Marg und Partner
Page 154 top kindly provided by Von Gerkan, Marg und Partner
Page 154 bottom kindly provided by Von Gerkan, Marg und Partner
Pages 154-155 kindly provided by Von Gerkan, Marg und Partner
Page 155 top kindly provided by Von Gerkan, Marg und Partner
Page 115 bottom kindly provided by Von Gerkan, Marg und Partner
Page 156 top kindly provided by Von Gerkan, Marg und Partner

Page 156 bottom kindly provided by Von Gerkan, Marg und Partner
Pages 156-157 kindly provided by Von Gerkan, Marg und Partner
Page 157 kindly provided by Von Gerkan, Marg und Partner
Page 158 Cyril Delettre/REA/Contrasto
Page 159 kindly provided by Marc Mimram Ingenierie S.A.
Pages 160-161 kindly provided by Marc Mimram Ingenierie S.A.
Page 161 top Roger Viollet/Alinari
Page 161 bottom kindly provided by Marc Mimram Ingenierie S.A.
Pages 162-163 Peter Cook/VIEW
Page 163 Alamy Images
Page 164 Dennis Gilbert/VIEW
Pages 164-165 Dennis Gilbert/VIEW
Page 166 Graeme Peacock
Pages 166-167 Graeme Peacock
Page 168 top kindly provided by Wilkinson Eyre Architects Limited
Page 168 center kindly provided by Wilkinson Eyre Architects Limited
Page 168 bottom kindly provided by Wilkinson Eyre Architects Limited
Page 169 top kindly provided by Wilkinson Eyre Architects Limited
Page 169 bottom kindly provided by Wilkinson Eyre Architects Limited
Pages 170-171 top Andrew Putler
Pages 170-171 bottom kindly provided by Wilkinson Eyre Architects Limited
Page 172 top Graeme Peacock
Page 172 bottom Graeme Peacock
Pages 172-173 Alamy Images
Page 173 Graeme Peacock
Pages 174-175 Graeme Peacock
Pages 176-177 PICIMPACT/Corbis/Contrasto
Page 177 kindly provided by Santiago Calatrava S.A
Page 178 top kindly provided by Santiago Calatrava S.A
Page 178 center kindly provided by Santiago Calatrava S.A
Page 178 bottom kindly provided by Santiago Calatrava S.A
Page 179 top kindly provided by Santiago Calatrava S.A
Page 179 bottom left kindly provided by Santiago Calatrava S.A
Page 179 bottom right kindly provided by Santiago Calatrava S.A

All the drawings of the bridges are by Angelo Colombo/Archivio White Star: pages: 3, 29, 31, 32, 36, 41, 49, 53, 55, 58, 61, 64, 70, 72, 77, 78, 84, 92, 99, 107, 109, 115, 118, 120, 124,129, 131, 136, 143, 147, 154, 161, 164, 172, 179.

AKNOWLEDGEMENTS

The publisher would like to thank the following:
Santiago Calatrava S.A
Arenas & Asociados
Honshu Shikoku Bridge Autority
Gleitbau Ges.m.b.H
Dissing+Weitling
Øresunsbro Konsortiet
Studio Carlos Fernandez Casado
Marc Mimram Ingenierie S.A
Von Gerkan, Marg und Partner
Wilkinson Eyre Architects Limited